The Technique of Parents Innovation and Independent Parents Cultivation in Sugarcane Cross Breeding

Edited by Wu Caiwen, Liu Xinlong, Hu Xin, Zhao Peifang

 China Agricultural Science and Technology Press

图书在版编目（CIP）数据

甘蔗亲本创新及独立亲本系统培育技术 = The Technigque of Parents Innovation and Independent Parents Cultivation in Sugarcane Cross Breeding / 吴才文等著. --北京：中国农业科学技术出版社，2022.8
ISBN 978-7-5116-5856-2

Ⅰ.①甘… Ⅱ.①吴… Ⅲ.①甘蔗－栽培技术 Ⅳ.①S566.1

中国版本图书馆CIP数据核字（2022）第139019号

责任编辑　周丽丽
责任校对　李向荣
责任印制　姜义伟　王思文

出 版 者	中国农业科学技术出版社
	北京市中关村南大街12号　邮编：100081
电　　话	（010）82109194（编辑室）　（010）82109702（发行部）
	（010）82109709（读者服务部）
网　　址	https://castp.caas.cn
经 销 者	各地新华书店
印 刷 者	北京建宏印刷有限公司
开　　本	170 mm×240 mm　1/16
印　　张	9.25
字　　数	300千字
版　　次	2022年8月第1版　2022年8月第1次印刷
定　　价	60.00元

◀ 版权所有·侵权必究 ▶

The Technigque of Parents Innovation and Independent Parents Cultivation in Sugarcane Cross Breeding

Contributors:

Wu Caiwen[1]　　Liu Xinlong[1]　　Hu Xin[1]　　Zhao Peifang[1]
Su Huosheng[1]　Wu Jinyu[1]　　　Li Yuxuan[1]　Yao Li[1]
Xia Hongming[1]　Qin Wei[1]　　　Zhao Jun[1]　　Liu Jiayong[1]
Zan Fenggang[1]　Zhao LiPing[1]　 Wu Zhuandi[1]　Zhu Jianrong[1]
Yang Kun[1]　　　Zhao Yong[1]　　Chen Xuekuan[1]
Eid Mohamed Eid Mehareb[2]　　Fouz Fotouh Mohamed Abo-Elenen[2]

---

[1] Sugarcane Research Institute, Yunnan Academy of Agricultural Sciences (YAAS); Yunnan Key Laboratory of Sugarcane Genetic Improvement.
[2] Sugar Crops Research Institute (SRCI), Agriculture Research Center (ARC); Egypt.

Preface

Parents are the material basis of sugarcane breeding. Only the breakthrough of parents can breed breakthrough sugarcane varieties. Research and utilization of sugarcane germplasm resources, continuous breeding to meet the needs of production of fine varieties, variety improvement and renewal are the fundamental guarantee for the sustainable development of the cane sugar industry. Since sugarcane sexual hybridization breeding, the genetic basis of sugarcane breeding worldwide can be traced back to 8 *S. officinarum*, 1 *S. barberi,* and 2 *S. spontaneum* cultivated and utilized by parents of "POJ" and "Co" systems. The long-term and high-frequency use of the 2 parent systems has led to the narrow genetic basis of sugarcane parents, the serious phenomenon of network and close relatives.

Relying on Sugarcane Research Institute, Yunnan Academy of Agricultural Sciences (YSRI), Yunnan has established the only National Germplasm Repository of Sugarcane in China, which collects and preserves 15 species of 6 genera from 34 countries and 12 provinces in China, 4,152 germplasm resources have been cataloged. The number of germplasm resources is second only to the world's largest Sugarcane Germplasm Resource Conservation Center, in which 5,743 germplasm resources are preserved, in India. Why are there many germplasm resources, and the research conditions are getting better and better? Still, the problems of genetic basis narrow, blood network, and kinship have long puzzled the sugarcane field. The results showed that: Due to different basic hybridization methods, the contribution to sugarcane breeding was completely different. Screening excellent original sugarcane species, breeding new independent parent system using peer to peer hybridization, good cross symmetry, clear blood relationship, the whole pedigree diagram is an inverted triangle, strong heterosis of progeny, as parents, breeding more varieties, more breakthrough varieties of progeny, a great contribution to breeding. It could breed excellent variety, the great breakthrough of variety traits, large promotion area, and great contribution to the industry, such as POJ2878, Co290, Co419, CP49-50,

CGT11, and CYZ81-173.

In order to meet the needs of the development of cane sugar industry, meet the needs of sugarcane science and technology workers to understand and master the dynamics of the creation of the new independent parent system, improve the breeding efficiency of sugarcane germplasm resources, transform resource advantages into breeding advantages and industrial advantages, and make a good turn of the sugarcane seed industry, we have written this book *"The Technique of parents Innovation and Independent Parents Cultivation in Sugarcane Cross Breeding"*. The writing of this book was supported by the China Agriculture Research System of MOF and MARA (CARS- 170101), Yunnan Sugarcane Germplasm Innovation And New Variety Breeding Team (2019HC013), and the Sugar Industry Science and Technology Mission of GengMa County, Yunnan Province.

The book was divided into five chapters, namely, origin and taxonomic status of saccharum species, the contribution and value analysis of sugarcane and its wild Resources in hybrid breeding, the hybrid mode and effect of sugarcane wild resources, the hybrid mode and effect of original saccharum cultivated species, cultivation methods of improvement, innovation and independent parent system in sugarcane hybrid breeding.

It is true that, due to the limited level of editors, the omissions in the book are inevitable; some views are relatively new, the understanding among peers may not be the same; in the actual analysis also exist a small number of consanguinity of parents is not very good, but offspring many varieties of individual cases, as long as the development of the cane sugar industry is beneficial in the field of sugarcane hybrid breeding continue to find unknown, untied unknown, for the truth and explore, for academic and contention is worth it, ask readers not hesitate to correct, to improve the together further to improve the efficiency of sugarcane hybrid breeding to make more enormous contributions.

Editors

June 2021

Contents

1 Origin and Taxonomic Status of Saccharum Species 1

   1.1 The Origin of Sugarcane 1

   1.2 The Systematic Classification and Evolution of Sugarcane Germplasm Resources 5

   References 10

2 The Contribution and Value Analysis of Sugarcane and its Wild Resources in Hybrid Breeding 14

   2.1 The Breeding Contribution of Original *Saccharum* Species 14

   2.2 The Characteristics and Breeding Contribution of Wild *Saccharum* Species 18

   2.3 The Relatives and Breeding Value Analysis of Genera Related to *Saccharum* 22

   References 31

3 The Hybrid Mode and Effect of Sugarcane Wild Resources 33

   3.1 Ways and Effects of Hybrid Utilization of Wild Species of Related Genus of Sugarcane 34

   3.2 The Hybrid Utilization Mode and Effect of Wild Saccharum Species 49

   References 62

4 The Hybrid Mode and Effect of Original Saccharum Cultivated Species 67

   4.1 The Hybrid Utilization Mode and Effect of *S.officinarum* 67

   4.2 The Hybrid Utilization Mode and Effect of *S.barberi* 77

   4.3 The Way and Effect of Hybrid Utilization of *S.sinense* 78

4.4 Ways and Effects of Interspecific Hybridization for Original
    Saccharum Cultivated Species 81
References 82

5 Cultivation Methods of Improvement, Innovation and Independent Parent System in Sugarcane Hybrid Breeding 85

5.1 Independent Parents Breeding and Its Contribution to Sugarcane
    Breeding 86
5.2 Cultivation of Innovative Sugarcane Parents and Their Contribution
    to Sugarcane Breeding 98
5.3 Breeding Improved Sugarcane Parents and Their Contribution to
    Sugarcane Breeding 103
5.4 Influence of Hybridization Mode on the Traits of Offspring 125
5.5 Advances in Breeding New Independent Parents of Sugarcane by
    the Peer to Peer Hybridization in Yunnan 127
References 135

1 Origin and Taxonomic Status of Saccharum Species

To survive, ancient humans mainly collected wild animals and plants to obtain food. With the long-term use of wild plants and the further understanding of the growth habits of wild plants, the cultivation techniques are constantly improved to create conditions suitable for the growth and development of plants, and the selection and cultivation are carried out at the same time. The research on the origin and classification of sugarcane resources can help scientific and technological workers to understand the genetic resources of sugarcane genera and species, establish "gene bank", use useful genes to transform existing varieties, and breed new varieties for people to use; At the same time, through the understanding of the ecological and geographical conditions of the origin of crops, we can achieve the purpose of artificial controlling the growth of crops. Therefore, the study on the origin and taxonomic status of sugarcane species is conducive to improving resource utilization efficiency, cultivating more new varieties of sugarcane, and contributing to the development of the sugarcane industry.

1.1 The Origin of Sugarcane

According to textual research, humans have cultivated sugarcane for more than 4,000 years (Chen et al., 2011). The question about the origin of cultivated sugarcane remains unclear now; many opinions from scientists in different regions or counties keep the argument. Some say it originated in China, some say it originated in New Guinea, and some say it originated in India and Southeast Asia. The history of cultivating sugarcane in China is very early, China is classified as one of the oldest sugarcane growing countries. Sugarcane recorded firstly in ancient Chinese documents appeared in an old book named as "Chu Ci" written by Qu Yuan, in the Warring States Period at the end of the 4th century BC, the word "Zhe" or "Shu Zhe" was firstly found in the literature included in "Chu Ci·Zhao Hun Fu". "Zhe" is the oldest name for Sugarcane, and "Zhe Jiang" is a concentrated product of sugarcane juice. According to the record from "Cane Creation Study" written by Wang (1979), and the book about Chinese agriculture written by Wagner from German, proposed that Zuckerrohr had been called sugarcane since the most ancient times, This indicates that sugarcane originated really in China. According to the recorded document above, the conclusion is that sugarcane grown in China is an inherent crop for China. G.R. Porter emphasized that China was the first country to grow sugarcane in his book named "History of Sugarcane"; this opinion was also supported by the reference named "A Brief History of Sugarcane in China and the Origin of Sugarcane" written by Zhou Keyong. Based on the eight origin centers of cultivated plants in the world proposed by the world-renowned botanist and geneticist H.S.Vavilov, Zhou Keyong proposed that China is one of the original centers of sugarcane. He once asked Bao Wenkui to ask Indian experts about the difference in sugarcane names between India and China when Bao Wenkui went to India to attend a genetics conference in 1982. The Indian expert replied that white sugar was called "Chent", a homophone with "China", in India, this indicated that sugar technology was introduced into India from China. However, according to the book named "Sugarcane Science" written by Luo JS, sugar production technology originated in India. India should be the origin place of sugarcane. For the opinion, being supported not only by these records from some Indian literature traced back two to three thousand years but also by these facts that China sugarcane was introduced from India and lots of wild sugarcane named *Saccharum spontaneum* L. distributed widely in these Indian regions from Bangladesh to Assam and Sikkim. But Wang (1979) proposed that this evidence supporting the opinion was insufficient because the wild sugarcane also grew in other regions like China and Southeast Asia. There was a wide distribution range for wild sugarcane, ranging from west to Turcomania and Afghanistan, east to Melanesia, Taiwan Province of China, and other places.

The opinion that sugarcane originated in Papua New Guinea was a deduction from American Brandes after collecting wild Sugarcane in Papua New Guinea in the late 1920s most authoritative sugarcane expert in the United States and had a significant influence. For this view, some people agree, some people oppose it. Wang Qizhu thought that Brandes's opinion is reasonable to a certain extent. For the origin of sugarcane, the critical question was which type of sugarcane was considered. So it isn't easy to gain a consistent conclusion based on different sugarcane types. Wang Jianming proposed in his book "The Origin of Sugarcane" that *Saccharum officinarum* originated in Southeast Asia was accepted by lots of people before Brandes's opinion was issued. It is also believed that *S. officinarum*, *Saccharum robustum,* and *S. spontaneum* from Borneo and Sulawesi were planted in Papua New Guinea by early human activities. Due to the complex terrain and various climate types in Papua New Guinea, many new types were produced so that Brandes can collect more rich types of sugarcane germplasm containing *S. officinarum*, *S. robustum*, *S. spontaneum,* and *Saccharum edule.* Therefore, it was inappropriate that Papua New Guinea was considered the only place of origin for *S. officinarum* and *S. robustum*. It should be emphasized that Borneo and Sulawesi were the origin place of these sugarcane species.

For naming the rule of sugarcane, without adding these words like "Hu" "Fan" "Yang", foreign names, and transliteration used naming imported plants, the name of sugarcane was purely based on the habit of domesticated plants in China. The relative sugarcane species are widely distributed in China, and there are pretty rich types involving nine genera of the sugarcane subfamily. *S. spontaneum* is distributed widely in North to Qinling mountains and south to Hainan island among these sugarcane germplasm resources, especially many clones distributed in south China and South Southwest China. By analyzing the isomerase and flavonoids of sugarcane to determine the genetic relationship among sugarcane species, Australian scientists proposed that sugarcane ancestors were *S. spontaneum* and *Miscanthus flordulus*, southern and southwestern of China were important distribution center of two wild species. According to these results mentioned above, the south and southwestern of China are some of the original places of sugarcane. *Saccharum sinense* Roxb from the *Saccharum* genus originated in China.

In the 1980s, Chinese scientific research institutes jointly conducted wild sugarcane surveys and proved that Guangdong, Guangxi, Fujian, Yunnan, Guizhou, Sichuan, and Tibet have wild sugarcane distribution. Especially in the south of Yunnan province, there are more wide distribution and genetic diversity for wild sugarcane. Yang et al. (1996) observed and calculated the chromosome type of 87 clones of *S. spontaneum* from Yunnan, found four types which were $2n$=60, 64, 70, 80. By investigating these clones of *S. spontaneum* from different regions of China, Wen et al. (2001) found that the types of the chromosome from China clones were wealthy, including 11 chromosome types like $2n$=60, 64, 70, 72, 78, 80, 90, 92, 96, 104, 108, among which the three types ($2n$=64, 80, 96) showed the highest distributing frequency, and $2n$=104 and $2n$=108 were firstly reported in China. 25 random amplified polymorphism DNA (RAPD) selected by Fan et al. (2001) was used to investigate 82 clones of *S. spontaneum* from the different populations in Yunnan and four *S. spontaneum* clones from foreign. The results showed that *S. spontaneum* in Yunnan possesses abundant genetic diversity. The genetic diversity of low latitude types is significantly higher than that of high latitude types. The polymorphism was less and less when latitude and elevation increased. Based on DNA data analysis results,86 clones of *S. spontaneum* were divided into eight different ecotypes. The topology of the phylogenetic tree of *S. spontaneum* corresponds to their geographical distribution. The initial results indicated that *S. spontaneum* in Yunnan was probably originated from the south of Yunnan, then diffused to high elevation and high latitude of the northwest and the northeast, so firstly proposed that southern Yunnan may be one of the original centers of wild sugarcane. Zhou Keyong also cited many facts in his article "A brief history of cane sugar in China and the origin of sugarcane" and criticized the view that sugar-making technology of sugarcane in China was imported from India. At the same time, it is also one-sided to claim that sugarcane only originated in India. For the view that Tang Xuanzang introduced Indian sugar technology to China and established China's first sugar industry in Neijiang, Sichuan, when he went to India to learn from the scriptures in the third year of Zhenguan in the Tang Dynasty (629 AD), and returned to Xian in 645 AD to through Tibet, Chengdu, Neijiang. Zhou Keyong thought the view was filled with various contradictions. The record of sugarcane plantations in China can trace back to the 4th century BC, according to "ChuCi·ZhaoHunFu" wrote by Qu Yuan.

Based on textual research on language and documentation, investigation, collection, researches on the types and distribution of wild sugarcane, and the growth and development of cultivated sugarcane, most scholars suggested that the origin of sugarcane may be in India, Vietnam, China, and the Pacific islands near Southeast Asia.

1.2 The Systematic Classification and Evolution of Sugarcane Germplasm Resources

1.2.1 Systematic Classification of Germplasm Resources of *Saccharum* L.

The classification of *Saccharum* L. and its relatives is shown in Figure 1-1.

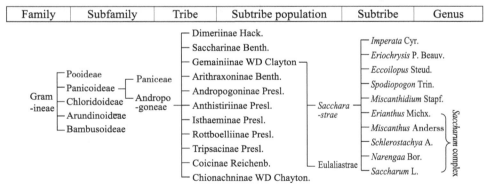

Figure 1-1 The status of *Saccharum* complex in Gramineae (Daniels et al., 1987)

Sugarcane (*Saccharum* spp.) was classified in *Saccharum* L., Saccharinae, Andropogoneae, Panicoideae, Gramineae, Glumiflorae, Monocotyledoneae, Spermatophyta. In Saccharinae, some important exploiting potential species have closely related with these species of *Saccharum*. The Indian scholar collected and studied Indian wild sugarcane and its relatives, and Mukherjee (1957), who was an

Indian scholar, carried out the collection and research of wild sugarcane resources and proved that some genera related to the origin of sugarcane like *Saccharum* L., *Erianthus* Michx., *Sclerostachya* A., *Sclerostachya* A., and Narenga Bor. have a close relationship with sugarcane, and can cross with sugarcane, finally, he invented a new term of "*Saccharum* Complex" to represent this kind of germplasm resources as a more extensive "breeding pool" in 1957. Based on the characteristics of this "breeding pool", Daniels and Daniels (1975) suggested that *Miscanthus* Anderss should also be included in "*Saccharum* Complex". Besides, Wen (1998) thought that these species from *Imperata* and *Sorghum* should be put into the "breeding pool" because of their close relationships with *Saccharum*.

1.2.2 Systematic Evolution of the Germplasm Resources of Sugarcane

For *Saccharum*, there are two wild species and four original cultivated species, of which two wild species were called large-stemmed wild species and thin-stemmed wild species whose Latin names were *S. robustum* and *S. spontaneum*, respectively. Four original cultivated species were named tropical species (*S. officinarum*), Indian species (*S. barberi*), Chinese species (*S. sinense*), and succulent flower spike wild species (*S. edule*). *The domestication and evolution of S. robustum* directly form *S. officinarum*. *S. barberi* and *S. sinense* were confirmed as natural hybrids of *S. officinarum* and native *S. spontaneum*. However, the origin of the ear-eating species remained unclear. The systematic classification and evolutionary research of sugarcane germplasm resources not only revealed the evolutionary relationship between sugarcane and its related genera but also provided a theoretical basis for broadening the genetic basis of sugarcane, and then clarified several problems in the classification and evolution of sugarcane genus and species, finally, do help to clarify the utilizing purpose of sugarcane germplasm clearly and significantly improve exploiting efficiency.

Price (1967) reported that the chromosome number of the *S. robustum* was 80 and relatively stable, and the tropical species (*S. officinarum*) possessing considerable morphological variation also had 80 chromosomes mostly belonging to bivalent. The studies of Barber (1916) and Jeswiet (1927) showed that Indian species (*S. barberi*) had seven kinds of chromosome types which were $2n$=81, 82, 83, 91, 92, 105, 119, and the range of chromosome number of Chinese species (*S. sinense*) was between 106 and 120. Luo (1984) thought that *S. barberi* and *S. sinense* are natural hybrids between the tropical and *S. barberi* based on the analysis of morphology and somatic cell chromosome number. By comparing cytoplasmic DNA and the cluster analysis of nuclear DNA, Irvine (1999) concluded that *Saccharum* species should be divided into two main clusters: *S. spontaneum* cluster and *S. officinarum* cluster other species and interspecific hybrids. Irvine also agrees that the *S. barberi* and the *S. sinense* may be natural hybrids between *S. officinarum* and *S. spontaneum*. D'Hont et al. (1993) found no differences in chloroplast genomes among these species used heterogeneous probes to study the cytoplasmic diversity within *Saccharum*. These differences showed in mitochondrial genomes belonging to maternal inheritance were consistent with taxonomic conclusions and displayed remarkable diversity among these clones of *S. spontaneum*. According to RFLP marker data, Lu et al. (1994) divided 50 accessions of *Saccharum* germplasms into three groups using 94 maize single-copy probes named thin-stemmed wild, *S. robustum* group, and *S. officinarum* group. More rich genetic diversity appeared among *S. spontaneum* groups; it also demonstrated that the *S. sinense* and the *S. barberi* might be hybrids of *S. officinarum* and *S. spontaneum*, consistent with Luo's view (1984). By further analyzing genetic similarity among 40 interspecific hybrids, Lu et al. found that these genetic differences of hybrids are mainly determined by the genetic basis of *S. spontaneum* with a 0.61 similar similarity coefficient with the *S. officinarum*.

Cai et al. (2005) thought that the thin stem specie was far from other species. More close relationships were found between the cultivated varieties and the *S. officinarum* or the *S. sinense* and the *S. barberi*. According to AFLP marker data, *S. robustum* clustered together. After finishing the band variation analysis of esterase (EST) and peroxidase (POD) isoenzyme among *S. spontaneum* clones from Fujian,

Yunnan, Sichuan, Guangxi, and other collecting regions. Wang et al. (1997) and Qiu et al. (1987) found that these clones from low altitude regions possessed more primitive bands and formed a trend that the enzyme bands of the population of original location gradually spread to the population of high altitude region. The results showed that the *S. spontaneum* developed from low-altitude regions to high altitude regions. The results from the report of Fan et al. (2001) pointed that these clones of *S. spontaneum* from different ecological regions in Yunnan appeared prominent geographical distribution characteristics and preliminary proved that the *S. spontaneum* might originate from low altitude and low altitude regions in southern Yunnan, then expanded to the northwest and northeast regions of high altitude and high latitude so that southern Yunnan may be one of the original centers of the *S. spontaneum*. By analyzing the genetic relationships within *Saccharum* germplasms, Nair et al. (1999) concluded that the *S. officinarum* had low genetic diversity and had a close relationship with the *S. robustum*, which may be the ancestors of *S. officinarum*, while a distant relationship with the *S. spontaneum*.

For the relationships among *Saccharum* complex, these clones from *Saccharum* complex were divided into eight taxa by Nair et al. (1999) according to RAPD marker data, which were *S. officinarum*, *S. robustum*, *S. spontaneum*, S.barberi, *S. sinense*, *Erianthus*, *Sclerostachya,* and *Narenga,* the close relationship between *Sclerostachya* and *Saccharum*. *Erianthus* had a high degree of divergent evolution with *Saccharum*. Cai et al. (2005) thought that the relationship between the *S. spontaneum* and *Erianthus* was more close to the relationship between the *S. officinarum* and *Erianthus rufipilus*, *Saccharum* was closer to *Miscanthus* than *Narenga* and *Erianthus*, and *Narenga* had a close relationship with *Erianthus*; for *Erianthus arundinaceum* was more closely related to *Saccharum*, followed by *Miscanthus,* while a further relationship with *Narenga* and *Erianthus*; for *Erianthus rockii,* had a more close relationship with *Narenga*, follow by *Erianthus*, and shows a distant relationship with *Saccharum* and *Miscanthus*. Regarding the taxonomic status of *Erianthus arundinaceum*, Feng et al.(1997) and Cai et al. (2005) believed that this species should be included in the *Saccharum* genus or put into a new genus based on molecular marker data.

Among the related genera of *Saccharum*, much research on *Erianthus* resources due to their excellent flowering habit, strong rationality, growth vigor, and stress resistance has been widely exploited in breeding programs. Sawazaki et al. (1989) and Glaszmann et al. (1989) demonstrated different clones of *Saccharum* Complex had different isoenzymes in peroxidase (POD), which can reflect the genetic relationships among different clones. Xiao et al. (1994) used POD isoenzyme electropulse technique to compare the relationships between *Erianthus arundinaceum* and *Saccharum* or *Erianthus*; the results showed that it is reasonable to classify *Erianthus arundinaceum* into *Erianthus*. Besse et al. (1996) found that the most extensive genetic distance within *Erianthus* appeared between two clones with chromosome type of $2n=20$ and other germplasms resources for 65 clones from *Erianthus* and 14 clones from *Saccharum* with a mean genetic distance of 0.748 using RFLP marker data from 14 nuclear single copy probes, and found a unique marker that can be used to identify *Erianthus* resources. And the genetic relationships between *S. spontaneum* and *Erianthus* are more close than those between *S. officinarum* and *Erianthus*. In 1999, Besse and McIntyre experimented with the fluorescence and non-fluorescence in situ hybridization of *Erianthus* elephantinus ($2n=20$) *procerus* ($2n=40$) using wheat nuclear DNA probes, 2 and 4 rDNA sites found respectively indicated that the chromosome base is 10 for two species. This result was consistent with the clone studies with a chromosome number of $2n=60$ reported by Hont et al. (1996).

Besse et al. (1996) compared the differences between the 18S+26S and 5S rDNA loci in 62 clones belonging to 11 species of *Erianthus* and 15 clones from 2 species of *Saccharum,* some differences between genera found in the length of 5S units and restriction sites can be used to illustrate the evolutionary relationships between these species of *Erianthus*, finally obtaining a unique marker specific to *Erianthus* which was an extra *BamHI* cut site comparing to *Saccharum,* the markers help detect the introgression in offspring from two genera. *E.trinii* and *E.brevibardis* had a very close relationship with *S. spontaneum* and *Miscanthus Sinensis*. By comparing the rDNA restriction maps of Sugarcane, species of *Erianthus*, sorghum, and maize, the relationships between *Erianthus* and other genera were further clarified, and the classification position of *Erianthus rufipilus* in *Saccharum* complex was determined clearly.

In summary, with the continuous improvement of the theory of sugarcane genetic evolution, the update and development of detecting equipment and technology, there will be a clearer understanding of the genetics and evolutionary mechanisms of sugarcane for all researchers, that will further promote the development of sugarcane genetic improvement and accelerate the progress of sugarcane industry.

References

ADATI S, SHIOTANI I, 1962. The cytotaxonomy of the genus *Miscanthus* and its phylogenetic status[J]. Bulletin of the Faculty of Agriculture Mie University (25): 1-24.

BARBER C A, 1916. Studies in India sugarcane, II. Memoirs: Department of Agriculture in India[J]. Botanical Series (8):103-199.

BESSE P, MECINTYRE C L, BERDING N, 1996. Ribosomal DNA variations in *Erianthus*, a wild sugarcane relative (Andropogoneae-Saccharinae)[J]. Theoretical and applied genetics, 92 (6):733-743.

BESSE P, TAYLOR G, CARROLL B, et al, 1998. Assessing genetic diversity in a sugarcane germplasm collection using an automated AFLP analysis[J]. Genetica, 104 (2):143-153.

BURNER D M, PAN Y B, WEBSTER R D, 1997. Genetic diversity of North American and Old World *Saccharum* assessed by RAPD analysis[J]. Genetic Resources and Crop Evolution, 44 (3):235-240.

CAI Q, FAN Y H, AITKEN K, et al., 2005. Assessment of the Phylogenetic Relationships within the "*Saccharum* Complex" Using AFLP Markers[J]. Acta Agronomica Sinica, 31(5): 551-559.

CAI Q, WEN J C, FAN Y H, et al., 2002. Chromosome analysis of *Saccharum* L.

and related plants[J]. Southwest China Journal of Agricultural Sciences, 15 (2): 16-19.

CHEN N W, YANG R Z, WU C W, et al., 1995. A study on isoenzymes of esterase of wild germplasm plants related to sugarcane in Sichuan[J]. Sugarcane, 2 (1):7-13.

CHEN R K, XU L P, LIN Y Q, et al., 2011. Modern sugarcane genetic breeding[M]. Beijing: China agricultural press (in Chinese).

CHEN S F, HE J, ZHOU P H, et al., 2008. The Karyorypes of M*icanthus sinensis* and *M. floridulus*[J]. Acta Agriculturae Universitis Jiangxiensis (2) :123-126.

CORDEIRO G M, CASU R, MCINTYRE C L, et al., 2001. Microsatellite markers from sugarcane ESTs cross transferable to *Erianthus* and sorghum[J]. Plant Science, 60 (6):1115-1123.

CORDEIRO G M, PAN Y B, HENRY R J, 2003. Sugarcane microsatellites for the assessment of genetic diversity in sugarcane germplasm[J]. Plant Science, 165 (1):181-189.

DANIELS J, ROACH B T, 1987a. A taxonomic listing of *Saccharum* and related genera[J]. sugarcane Cane, Spring (Supplement):16-20.

DANIELS J, ROACH B T, 1987b. Taxonomy and evolution[M]// HEINZ D J. Sugarcane Improvement Though Breeding. Amsterda: Elsevier.

DANIELS J, DANIELS C A, 1975. Geographical, historical and cultural aspects of the origin of the indian and Chinese sugarcanes *S. barberi* and *S. sinense*[J]. Sugarcane Breeding Newsletters (36):4-23.

D'HONT A, GRIVET, FELDMANN P, et al., 1996. Characterisation of the double genome structure of modern sugarcane cultivars (*Saccharum* spp.) by molecular cytogenetics[J]. Molecular and General Genetics (250):405-413.

D'HONT A, LU Y H, FELDMANN P, 1993. Cytoplasmic diversity in sugarcane revealed by heterologous probes[J]. Sugarcane (1):12-15.

DUTT N L, RAO J T, 1950. The present taxonomic position of *Saccharum* and its congeners[J]. Proc. ISSCT (6):286-293.

FAN Y H, CHEN H, SHI X W, et al., 2001. RAPD analysis of *Saccharum spontaneum* from different ecospecific colonies in Yunnan[J]. Acta Botanica Yunnanica, 23 (3):298-308.

FENG D, CHEN R J, 1997. The study on using random amplified polymorphic DNA(RAPD) in the classification of *Saccharum arundinaceum* Retz[J]. Genomics and Applied Biology, 16 (4):261-267.

GLASZMANN J C, FAUTRET A, NOYER J L, et al., 1989. Biochemical genetic markers in sugarcane[J]. Theor Appl Genet, 78 (4):537-543.

GRASSL C O, 1972. Taxonomy of *Saccharum* relatives:*Sclerostachya*, *Narenga* and *Erianthus*[J]. Proc. ISSCT (14):240-248.

HUNTER A W S, 1934. A karyosystematic investigation in the Gramineae[J]. Canadian Journal of Research (11):213-241.

IRVINE J E, 1999. *Saccharum* species as horticultural classes[J]. Theor Appl Genet, (98):186-194.

LI X W, LUO J S, 1948. Cell studies on sugarcane plants - noble species, thatched grass and wild hybrids[J]. Botanical report of the National Central Research Institute (2):147-160.

LUO J S, 1984. Sugarcane science[C]. Guangzhou: Guangdong sugarcane Society.

MARIATERASA D C, TREVOR R H, SUSANNE B, 2010. Chloroplast DNA markers (cpSSRs, SNPs) for *Miscanthus*, *Saccharum* and related grasses (Panicoideae, Poaceae)[J]. Mol Breeding (26):539-544.

MUKHERJEE S K, 1957. Origin and distribution of *Saccharum*[J]. Bot. Gaz., (119):55-61.

NAIR N V S, NAIR T V, SREENIVASAN, et al., 1999. Analysis of genetic diversity and phylogeny in *Saccharum* and related genera using RAPD markers[J]. Genetic Resources and Crop Evolution (46):73-79.

PENG S G, 1990. Sugarcane breeding[M]. Beijing: Agricultural Press (in Chinese).

QIU C L, HE S C, YANG S Q, 1987. Studies II on *Saccharum spontaneum* L. in Yunnan - Esterase isozyme[J]. Journal of Yunnan Agricultural University, 1 (1):80-83.

SAWAZAKI H E, SILVAROLLA M B, ALVAREZ R, 1989. Isoenzymic

characterization of sugar clonea and somaclonea[J]. Bragantia, 48 (1):1-11.

TAKAHASHI S, FURUKAWA T, ASANO T, et al., 2005. Very close relationship of the chloroplast genomes among *Saccharum* species[J]. Theoretical and Applied Genetics, 110 (8):1 523-1 529.

TATEOKA T, 1954. Karyotaxonomy in Poaceae II. Somatic chromosomes of some species[J]. Cytologia (19):317-328.

WANG Q Z, 1979. Sugarcane Farming[M]. Taibei: National editorial Gallery.

WANG S L, WANG S Q, GUO C F, et al., 1997.Study on esterase isozyme of Fusarium and bamboo cane in Fujian Province[J]. Sugarcane, 4 (1):9-13.

WEBSTER R D, SHAW R B, 1995. Taxonomy of the native North American species of *Saccharum* (Poaceae: Andropogoneae)[J]. Sida,16 (3):551-580.

WEN J C, CAI Q, FAN Y H, et al., 2001. Studies on the chromosome numbers of *Saccharum spontaneum* and related plants-*Erianthus arundinaceum*, *Narenga* in China[J]. Sugarcane and Canesugar (3):12-15.

WU C W, ZHAO P F, XIA H M, et al., 2014. Modern cross breeding and selection techniques in sugarcane[J]. BeiJing: Science Press (in Chinese).

XIAO F H, 1994. A comparative study on plant morphology and isozyme between *Erianthus arundinaceum*, Sugarcane and *Erianthus fulvus*[J]. Sugarcane, 1 (1):22-27.

XIAO F H, LI F S, 1996. A study on wild relative species of Sugarcane (*Erianthus fulvus*)[J]. Sugarcane (Fujian), 3 (2):1-6.

YANG L H, 2004. Studies on posterity of *Saccharum spontaneum* L. and *Erianthus arundinaceum* Retz. and *Erianthus rockii* Keng for resistance to smut in Yunnan, China[J]. Sugarcane, 11(1):10-14.

YANG Q H, HE S C, 1996. Chromosome number and geographical distribution of *Saccharum spontaneum* in Yunnan Province[J]. Sugarcane (1):10-13.

YU H, ZHAO N X, 2004. Geographical distribution of Saccharinae (Gramineae)[J]. Journal of Tropical and Subtropical Botany, 12 (1):29-35.

2 The Contribution and Value Analysis of Sugarcane and its Wild Resources in Hybrid Breeding

Crop germplasm resources, which are also called variety resources, genetic resources or gene resources, as an important part of biological resources, are the genetic material basis for the breeding program. They also play important roles in improving agricultural production capacity, ensuring food safety, and maintaining sustainable agricultural production development. Sugarcane germplasm resources contain all plant genetic resources used for sugarcane breeding plans, including original sugarcane cultivar species, landraces, hybrid varieties, intermediate breeding materials, wild species, related species, etc. These resources have become the material basis for sugarcane breeding and the material basis for studying the origin, evolution, and genetic diversity of sugarcane and its relatives. As we know, modern sugarcane varieties are bred from multiple crosses and backcross selections by using the few original cultivar species, wild species, and relative species. Due to only a tiny number of germplasm resources used in breeding plans, sugarcane varieties possessed a narrow genetic basis, hindering continued improvement in agronomic traits, disease resistance, and adaptability. For resolving this problem, sugarcane breeders need to continue to collect, identify, evaluate new precious sugarcane germplasm resources, and continuously overcome a series of problems such as difficulty in booting, flowering, and non-matching in the flowering period through developing new technological methods, successfully finishing in breeding better new varieties or breakthrough varieties with the help of new "Noble Breeding Theory" and independent parental systems creating new mutant lines. The final purpose is to promote the continuous development of the cane sugar industry.

2.1 The Breeding Contribution of Original *Saccharum* Species

Six species in total are included in *Saccharum* L., which are tropical species (*S.*

officinarum), Indian species (*S. barberi*), Chinese species (*S. sinense*), thin-stemmed wild species (*S. spontaneum*), large-stemmed wild species (*S. robustum*), and succulent flower spike wild species (*S. edule*). The first five species had played an important role in the sugarcane breeding plan. The genetic basis of sugarcane varieties worldwide came from the interspecific hybrids produced by crossing each other among 3 or 5 species like *S. officinarum, S. barberi, S. sinense, S. spontaneum, S. robustum*. Initially, to overcome the outbreak of *S. officinarum* disease, the resistant genes of *S. spontaneum* were introduced into *S. officinarum* through the crossing. Then the *S. officinarum* was backcrossed repeatedly with offspring to restore the high sugar characteristics of *S. officinarum*; new wild species were exploited continuously to improve disease resistance and adaptability of sugarcane varieties.

2.1.1 The Characteristics and Contribution to Sugarcane Breeding of Tropical Species (*S. officinarum*)

The tropical species (*S. officinarum*) originated in the South Pacific and Oceania, its chromosome number is $2n=80$. It is suitable for cultivation in tropical and subtropical areas with high temperatures and a lot of rain. Some important commercial traits of modern sugarcane varieties, like high sugar content and resistance to smut, came from the *S. officinarum*. It performs very well in some breeding traits, for example, medium and large stems, tall plants, vigorous growth, high yield, high sugar content, low fiber content, more cane juice, et al. Certainly there are also some defects like weak tillering ability, underdeveloped root system, poor resistance to barrenness, drought, and cold. Some *S. officinarum* have been utilized successfully in the breeding plans, which contain Badila, Bandjarmasin Hitan, Lahaina, Fiji, Black Cherbon, Crystalina, Loether, *S. mauritius*, Veillai, White Transparent, and Kaludai Boothan, etc., and had made significant contributions to improving sugarcane varieties. Currently, the main cultivated varieties in the world all contained their consanguinity. So far, lots of *S. officinarum* resources had been collected and conserved in different gardens worldwide. About 325 clones of *S. officinarum* were preserved in Miami sugarcane germplasm resources nursery in the United States, but only few were used in the breeding program. More *S. officinarum* resources will be collected and evaluated in commercial traits and stress resistance with the rapid development and improvement of molecular biotechnology and flowering hybridization technology. By performing basic crossing with a clear breeding target, QTL mapping of elite traits and cloning of excellent genes for *S. officinarum* resources will help broaden the genetic basis of modern sugarcane varieties and obtain breakthrough development in some commercial traits such as yield and sugar content.

2.1.2 The Characteristics and Contribution to Sugarcane Breeding of Indian Species (*S. barberi*)

Indian species (*S. barberi*) is one of the sugarcane cultivated species, mainly distributed in Punjab, Baisha, and Sami in India's Ganges River Basin and distributed in southern China. It is suitable for cultivation in subtropical and temperate regions and had some elite traits that are needed for breeding, for example, early maturity, high sugar content, high fiber content, drought tolerance, barren tolerance, tolerance to extensive cultivation, developed root system, multiple tillers, strong ratoons, high resistant to diseases such as wilt disease, red rot, and glue drop disease, moderately resistant to root rot. Still, *S. barberi* is susceptible to mosaic disease and smut. For *S. barberi*, the main problems affecting hybrid breeding efficiency are flowering and poor pollen development. The chromosome number of *S. barberi* ranges between 82 to 124, which Grassle (1972) thought to be a natural hybrid between *S. officinarum* and awn. Among *S. barberi*, Chunnee had made the most contribution to sugarcane breeding. It was a critical cultivar and had an essential impact on hybrid sugarcane breeding as parents. Some elite varieties such as POJ213, NCo310, and CP34-120 were produced using Chunnee as crossing parent. The Co series varieties in India and many varieties in the world are the descendants of Chunnee. Other *S.*

barberi have almost no contribution to sugarcane breeding due to their difficulty in flowering, poor pollen development, and low fertility.

2.1.3 The Characteristics and Contribution to Sugarcane Breeding of Chinese Species (*S. sinense*)

China is one of the first countries to cultivate sugarcane and make sugar. Chinese species (*S. sinense*) is mainly distributed in southern China, northern India, Iran, and Malaysia. It is also one of the sugarcane cultivated species suitable for cultivation in subtropical and temperate regions. It possesses some good characters for sugarcane breeding, which included strong resistance to stress, tolerance to extensive cultivation, drought tolerance, barren tolerance, cold tolerance, developed root system, strong tillering capability, substantial ratooning, early maturity, high sugar, high fiber content, suitable for industrial sugar production, immune to wilt disease, resistant to root rot and mosaic disease. But *S. sinense* also has many defects such as susceptibility to red rot and red pattern, susceptibility to smut for most clones, lateral buds germinating quickly, thick wax powder affecting clarification in making sugar. There are about 40 clones in China. The primary representative clones are Uba, Bamboo cane, and Lu cane. The chromosome number of *S. sinense* ranges between 116 to 118, and it was speculated by Grassl (1967) that it might be a natural hybrid between the *S. officinarum* and *Miscanthus sacchariflorus*. Some people thought that it might be a natural hybrid between *S. officinarum* and *S. spontaneum*. Based on the exploiting efficiency of *S. sinense* in the nearly 100 years of sugarcane hybrid breeding history, Wu caiwen ensured that the *S. sinense* Uba is a natural descendant from crossing between tropical and *S. spontaneum*. Another reason is that Uba can be used as a male parent or female parent and its offspring showed good fertility. However, bamboo cane and Lu cane performed difficult in flower capability, they can only be used as the female parent, and their hybrid offspring had not successfully developed new varieties. Therefore, it was speculated that bamboo cane and Lu cane should belong to the distant hybrid offspring of *S. officinarum*.

2.2 The Characteristics and Breeding Contribution of Wild *Saccharum* Species

Wild *Saccharum* species is an essential germplasm resource. Because it grows in the natural environment and withstood various natural disasters and environmental stress for a long time, strong resistance to biological and abiotic stress has formed, such as disease resistance, insect resistance, drought tolerance, cold tolerance, flood tolerance, salt-alkali tolerance, senescence tolerance, etc. These excellent traits can provide a necessary genetic basis for the improvement of cultivars.

2.2.1 The thin-stemmed wild species (*S. spontaneum*)

The *S. spontaneum* is also called *S. spontaneum*; thin-stemmed wild species, or sweetroot grass, is distributed widely and can be found between 40 degrees north latitude and 40 degrees south latitude. Various wild plant community types were found, such as dwarf community, stemless community, and large stem community with a stalk length range of 0.3–1.5 m and plant height range of 0.5–5 m. In addition to the enormous changes in morphological characteristics, the number of chromosomes also varies greatly, ranging from 40 to 128. Wen et al. (2001) counted the chromosome number of 106 clones of *S. spontaneum* from Yunnan, Sichuan, Fujian, Guangdong, Guizhou, and Jiangxi. These results showed that the diversity in the chromosome types was vibrant; 11 types in total were found, which are $2n=60, 64, 70, 72, 78, 80, 90, 92, 96, 104$, and 108, of which three kinds of types ($2n=64, 80, 96$) appear high distribution frequency. Qiu et al. (1987), Chen et al. (1995), and Wang et al. (1997) found that *S. spontaneum* also showed rich diversity in the esterase (EST) isozymes and peroxidase (POD) isozymes by analyzing these clones collected in Fujian, Yunnan, Sichuan, Guangxi, etc. The breed value for *S. spontaneum* is mainly manifested in extreme adaptation to adversity such as desert, marshes, and saline-alkali land on the seashore. Its many excellent traits, which are underground stems, well-developed roots, barren tolerant, drought tolerant, early growth, rapid growth, multiple tillers, substantial ratooning, early maturity, high fiber, early flowering, and easy flowering, etc., determined that *S. spontaneum* is the most critical wild resource in sugarcane breeding plans. However, it also has some shortcomings such as low sugar content, less juice, pipe, and pith, so overcoming these disadvantages in raising new elite varieties generally requires repeated crossings and backcrossings at least more than three generations. All main varieties contained the consanguinity of *S. spontaneum* of Java and Indians, China *S. spontaneum* from Hainan Yacheng, Hainan Lingshui, and Manhao Yunnan also was introduced into some new varieties bred in China by innovation plans. In the 1960s, some varieties possessing early maturing, high sugar, and cold tolerance, such as CP65-357, were developed in Louisiana State of the United States using *S. spontaneum* resources (Fanguy, 1974). The breeding organizations in Taiwan Province of china used *S. spontaneum* resources to raise new varieties like ROC[①]16, ROC23, ROC24, etc., in the middle and late 20th century (Chen, 2006). The Hainan Sugarcane Breeding Station in China had developed more than 30 new sugarcane varieties such as CYT 89-113, CTC 16, CGT24, etc., these *S. spontaneum* clones from Hainan Yacheng, Hainan Lingshui, and Yunnan 75-2-11 (Chen et al., 2010). Using the *S. spontaneum* clone called Yunnan Manhao, Yunnan Sugarcane Research Institute, Yunnan Academy of Agricultural Sciences developed a new variety named CYZ99-155 (An et al., 2007). According to the genealogy data of varieties, the cytoplasm of the currently cultivated sugarcane varieties only comes from a few *S. officinarum*. The resistance to disease and stress of varieties weaken gradually with the increase of planting years. With the continuous development of molecular marker technology and some new progress obtained in sugarcane cytoplasmic genetics research using SSR markers (Pan et al., 2004; Takahashi et al., 2005; Mariaterasa et al., 2010), sugarcane breeders had begun to use the cytoplasm of *S. spontaneum* to develop some new varieties with some breakthrough traits.

In China, two clones of *S. spontaneum*, Yacheng and Yunnan 75-2-11, have

---

① ROC 表示新台糖系列品种，全书同。

been exploited successfully (Zhang et al., 2006) and produced some elite hybrid parents CYC 58-47, CYC 58-43, and CYC 82-108. Using the elite hybrid parents continuously, some excellent sugarcane varieties or breeding materials were raised successfully, for example, successfully developing many elite hybrids such as CYZ 95-155 using these clones of *S. spontaneum* from Yunnan (An et al., 2007).

2.2.2 The large stemmed wild species (*S. robustum*)

The *S. robustum* is also called Irian wild species, perennial herb plant, Mainly distributed in irian Island, Borneo, New Great Britain, and Celebes Island. It can grow up to 10 m in areas with hot and humid climates and deep soil. The main breeding traits available for large-stalked wild species as primary wild resources used in breeding plans are strong vigor, hard cane stems like bamboo, strong wind resistance, strong drought tolerance, well-developed underground rhizomes, strong ratooning, and high fiber content. But this species also has some defects like low sugar content and susceptibility to some diseases such as mosaic, root rot, Fiji, and open fungus. Some breeding organizations in different countries used the species to cross with *S. officinarum*, then their offspring were backcrossed with commercial varieties and produced some excellent varieties. Since the strong wind resistance was found in PT43-52 belonging to the F_2 generation of *S. robustum* in Taiwan Province of China in 1946, a series of new varieties containing the consanguinity of *S. robustum* (Table 2-1) had been developed used PT43-52 and its derivatives as parents in breeding plans. Some progenies belonging to the F_3 generation of PT43-52, which included F140, F141, F146, F147, F148, F149, F150, F151, F152, F154, etc., not only been planted wildly but also exploited deeply continuously as parents, and developed two kinds of varieties named as F and ROC with the number of more than 40. Some varieties such as ROC10, ROC16, ROC20, ROC22, and ROC25 performed very well since they were applied in China's mainland regions in 2000 with more than 670,000 hm² planting area for many years in China. Simultaneously, more than 30 sugarcane varieties/materials were bred using these ROC varieties as parents in China.

Table 2-1 Taitang sugarcane varieties and their parents with PT43-52 consanguinity of *S. robustum*

No.	Variety	Crossing combination	Generation	No.	Variety	Crossing combination	generation
1	PT43-52	POJ2878×PT4-345	F_2	19	F162	F146×PT52-609	F_4
2	F140	PT43-52×CP34-19	F_3	20	F165	F146×F152	F_4
3	F141	PT46-150×PT43-52	F_3	21	F166	F146×F152	F_4
4	F146	NCo310×PT43-52	F_3	22	F168	NCo310×F152	F_4
5	F147	H32-8560×PT43-52	F_3	23	F169	F146×F145	F_4
6	F148	F138×PT43-52	F_3	24	F171	F146×F145	F_4
7	F149	PT43-52×CP34-79	F_3	25	F172	F153×F152	F_4
8	F150	NCo310×PT43-52	F_3	26	F173	F153×F162	F_4
9	F151	NCo310×PT43-52	F_3	27	F174	PT58-188×F152	F_4
10	F152	H32-8560×PT43-52	F_3	28	F175	PT38-184×F152	F_4
11	F154	PT51-213×PT43-52	F_4	29	ROC1	F146×CP58-48	F_4
12	F155	F141×CP29-116	F_4	30	ROC4	F146×F145	F_4
13	F156	F141×CP34-79	F_4	31	ROC5	F146×F152	F_4
14	F157	F146×PT51-1	F_4	32	ROC10	ROC5×F152	F_4
15	F158	PT52-50×CP34-79	F_4	33	F167	F156×F52-18	F_5
16	F159	NCo310×PT52-50	F_4	34	ROC2	F160×PT59-113	F_5
17	F160	NCo310×F141	F_4	35	ROC16	F171×74-575	F_5
18	F161	F146×PT52-227	F_4	36	ROC22	ROC5×69-435	F_5

Source: Wu et al., 2014.

China began the utilization of *S. robustum* germplasm resources in the 1970s. A new parent named CYC73-226 was produced using PT40-388 from the F_1 generation of *S. robustum* crossing with Co419. Then using CYC73-226 to raise some elite varieties such as CZZ79-177, CZZ80-101, CZZ92-126, CZZ93-213, CYT94-128, and CYT97-76.

The hybrids from offspring of *S. robustum* possessed many advantages, of which strong ratoon, drought tolerance, and good wind resistance were their most notable character. Most of the varieties selected from the offspring of *S. robustum* also showed some excellent features such as high germination rate, robust tiller, fast growth, high plant height, more stalks, good growth vigor and resistance to lodging, etc. There are also disadvantages like small and medium stems, low sugar content, late maturity, susceptibility to smut. Hence, it is essential to choose early maturing, high-sugar, and resistance to smut parents to cross with offspring of *S. robustum*.

2.2.3 The Succulent flower spike wild species (*S. edule*)

This species has a relatively small natural distribution range globally, mainly growing in some small islands located west of Irian and the Hebrides Islands. It was planted in the area from New Guinea to Fiji as a traditional vegetable of Malaysians. Its flow spike wrapped in leaf sheaths degenerated into thick fleshy-looking like corn ears. The main types of chromosomes are $2n=70-76, 84, 90-94$, etc. So far, the breeding value of this species has remained unclear.

2.3 The Relatives and Breeding Value Analysis of Genera Related to *Saccharum*

The species related to *Saccharum* are closely related to sugarcane and involved closely in the sugarcane breeding program. Modern sugarcane varieties are developed using a few *Saccharum* species such as *S. officinarum*, *S. barberi*, *S. sinense*, *S. robustum*, and *S. spontaneum* of the inner to cross each other. The led to a relatively high inbreeding coefficient and narrow genetic basis of modern sugarcane varieties, which had become the main bottleneck for current sugarcane breeding. Therefore, exploiting some new excellent germplasm resources from these genera related to *Saccharum* can broaden the genetic diversity of sugarcane varieties and create new variations based on "new noble breeding theory". Finally, improving breeding efficiency and developing more elite varieties have become the primary goal of all sugarcane breeders' efforts.

2.3.1 The Introduction of *Erianthus* and Its Breeding Value Analysis

Most of the *Erianthus* species belong to a tall and sturdy perennial herb with a chromosome type of $2n=20$, mainly distributed in temperate, subtropical, and tropical regions of Asia and Oceania. Generally, four species are considered the most prospective resources in *Erianthus* for sugarcane breed, which are *E. formosanus*, *E. arundinaceum*, *E. fulvus,* and *E. rockii*. At present, The two world sugarcane germplasm resource conservation centers in the United States and India have preserved this kind of precious resource. The National Germplasm Repository of Sugarcane has also conserved a certain amount of germplasm resources of *Erianthus* through years of hard work.

The main morphological characteristics of *Erianthus* species are as follows: long flat leaves, elongated panicles, spikelets with tufts at the base containing one degenerate sterile flower and one bisexual flower growing on each node (one is sessile, and the other is pedunculate), the main shaft broke down one by one after maturity, nearly equal glumes; the lemma with straight or twisted awns easily distinguished from *Saccharum*. Because these species of *Erianthus* possess some excellent characteristics like drought tolerance, barren tolerance, cold tolerance, waterlogging tolerance, clear leaves, strong resistance to disease, strong ratoon, high tillering capability, wide adaptability, etc., they were considered essential in sugarcane crossing breeding works by breeding workers of different countries.

Erianthus rufipilus is a wild species of *Erianthus* in Saccharanastrae. It mainly grows in tropical, subtropical, and temperate regions, mainly distributed in Yunnan, Guizhou, Sichuan, Hubei, Shaanxi, Tibet of China, northern India, Nepal, and Pakistan. In 1927, Ranke finished the crossing between *Erianthus rufipilus* and *S. officinarum* successfully. Later, Lagerfan (1953) reported that the chromosome transmission rule of the cross between *S. spontaneum* and *E. rufipilus* was in the way of $2n + n$. Xiao et al.(1992) found that the number of anthers varied from 0 to 3 anthers per spikelet, consistent with the variation range of each genus of

Saccharastrae and utterly different from the three anthers or two anthers per spikelet reported by predecessors. According to the measurement of brix of 7 lines in the natural growing environment in Yunnan, Tibet, Sichuan, and Shanxi of China, it was found that the brix value was generally high with an average of 12.9% and an enormous variation range of 3%–22%, and showed typical quantitative characteristics with a 34.5% of the coefficient of variation. Another three lines were performing a high Brix value of more than 20%. Yang et al. (1991) collected a clone of *E. rufipilus* in Shanxi, whose brix got 22%. It indicated that this species has the genetic basis of sucrose synthesis in stem similar to sugarcane. According to the available data of research reports and databases, these clones with more than 20% of brix were found only in these populations from *S. officinarum*, *S. sinense*, *S. barberi*, succulent flower spike wild species, *S. robustum*, *S. spontaneum*, and *E.rufipilus*. From the point of view of evolution, *E.rufipilus* is older than all *Saccharum* species, and it may be one of the ancestors providing high sugar genetic traits for sugarcane. The mainly distributed regions of *E.rufipilus* located on the border of India, Myanmar, and China, which is also the diversity origin center of *Saccharastrae*. Finally, some researchers proposed that *E.rufipilus* played a particular role in the origin and evolution of *Saccharum* by comparing this variation in chromosome number, karyotype, anther number, and brix, so it is worthy of further research and attention in breeding for *E.rufipilus*.

For *E.rockii*, Huang et al. (1997) had obtained a few seedlings from intergeneric hybridization between *E.rockii* as male parent and sugarcane cultivars, but these seedlings had a low mating ability. The Sugarcane Research Institute of Yunnan Academy of Agricultural Sciences (YSRI) carried out the utilization experiment of *E.rockii* and got a hybrid named CYZ 99-48 with superior resistance to both smut and leaf spot by crossing between CYR 91-2008 and CYR 93-3148 belonging to hybrid one progenies of *E.rockii* (Yang et al., 2004).

For *E.arundinaceum*, as early as 1885, Soltwedel began to cross between

E. arundinaceum and sugarcane but failed; in 1941, Jannacci obtained crossing progenies successfully between *S. officinarum* and *E.arundinaceum*, and the chromosome transmission of offspring was in the form of $n + n$. Cai et al. (2002) found three kinds of chromosome types in 46 clones of *E.arundinaceum*, $2n = 20, 40$, and 60, of which $2n = 20$ was the first report in China. In 1925, Jeswiet classified *E.arundinaceum* in *Erianthus*. Dutt and Rao confirmed this view in 1950. But Grassl (1972) suggested that seven species and six species not studied in *Erianthus* should be divided into a new genus named *Ripidium*. Many sugarcane colleagues have accepted this proposal. Still, because pteridophytes have long used the name of the new genus, the name was invalid. Daniels and Roach (1987) reclassified the genus as the Ripidium group (*Erianthus* sect. *Ripidium*), now the taxonomic status has been recognized internationally and has not been disputed overseas. However, the taxonomic status of *E.arundinaceum* in China was not unified (Xiao et al., 1996) because the six species in the Ripidium group discussed by Grassl were not observed in China. According to the literature records, there are four original species in *Erianthus*. There were significant differences between *E.rufipilus* and *E.arundinaceum* or *E.rockii*, so regarding them as representatives of *Erianthus* is insufficient in the taxonomic studies of *E.arundinaceum*. Webster et al. (1995) proposed the opposite view with Grassl, according to Clayton and Renvoize's report (1986), and believed that *E.rufipilus* should belong to *Saccharum* rather than *Erianthus*. Besse et al. (1998) pointed out that there were no close relationships among *Saccharum*, *Erianthus* sect. *Ripidium* and *Erianthus* from North America. In their relationships, *Saccharum* and *Erianthus* sect. *Ripidium* appeared more a close relationship; this was consistent with the "*Saccharum* complex" evolutionary pattern. Both *Saccharum* and *Erianthus* sect. *Ripidium* originated from the South Asian subcontinent. Giovani et al. (2001; 2003) ensured the view proposed (1998). Still inconsistent with Burner et al. (1997) obtained using RAPD marker data, *Saccharum* has a closer relationship with *Erianthus* of North America. Although some different views were issued on the taxonomic status of *E.arundinaceum*, it possessed many excellent characteristics needed in sugarcane breeding plans. *E.arundinaceum* has been considered necessary in innovation works by breeders from different countries, and some progress has been obtained.

2.3.2 The Introduction of *Miscanthus* and Its Breeding Value Analysis

Miscanthus Anderss was regarded as a tall perennial herb with thick and erect stems full of white soft pith and rhizomes. In *Miscanthus*, there are about 20 species globally, which are mainly distributed in these regions from Tahiti Island in the Pacific Ocean to Eastern Indonesia, Southeast Asia, Africa, Northern China, Northwestern Asia, and Japan. Their stalks can be used for making paper. About ten species were conserved in the National Germplasm Repository of Sugarcanein China. The stalk of these species from *Miscanthus* is very tall like *Phragmites communis* and small like *M. trandmorrisonensjs* with filamentous leaves. They have long linear-lanceolate leaves with a rough margin, white midvein, and a thick base. Comparing with *Saccharum* and *Erianthus*, *Miscanthus* has more strong rachis and pedicel spikelets. And it has more long setae and an awn in the fourth glume, which differs from sclerostachya. The inflorescence of *Miscanthus* belongs to a terminal panicle with a long raceme composed of slender racemes. Their spikelets with a short stipe and a long stipe have two bisexual flowers, pairs similar, their glumes densely covered with long hairs. *Miscanthus* has a particular economic value. It often was used as ornamental plant or fodder in Japan or the materials of building houses or archery equipment in Southeast Asia. *Miscanthus* has been redivided into four types: *Trierhena* Honda, *Eumiscanthus* Honda, *Kariyasua* Ohwi, and *Diandra* Keng (Peng, 1990; Zhang et al., 2006).

The chromosome number of *Miscanthus* ranges from 38 to 114. Chen et al. (2008) finished analyzing the karyotypes of two species of *Miscanthus* (*M.sinensis* and *M. floridulus*). The results showed that the chromosome numbers of both species were 38. The two species' karyotypes belong to 2B types with no satellite chromosome;

their karyotype formula was $2n = 38 = 28m + 10sm$ and $2n=38=24m+14sm$, respectively. Also, some chromosome type like $2n$ = 40, 42, 57, 60, 76, 95, and 114 were reported in some studies (hunter, 1934; Grassl, 1972; Daniels et al., 1975; Tateoka, 1954; Adati et al., 1962; Wen et al., 2001; Cai et al., 2002; Zhang et al., 2006).

Among the four types of *Miscanthus*, the representative species of *Trierrhena* Honda is *M. sacchriflorus* (Maxim) Benth, mainly distributed in Japan, central China, and Siberia. The common chromosome types is $2n = 76$, $2n = 57$ and $2n = 96$. This species has a root stem, bud, and root belt and possessed some morphological characteristics needed in sugarcane crossing breeding plans. Some scholars also considered it an important ancestor parent of *S. sinense* from the natural crossing offspring between *M. sacchriflorus* and *S. spontaneum*. For *Eumiccanthus* Honda, it included two species (*M. floridulus* (Lab.) Warb. ex Schum. et Laut. and *M. sinensis* Anderss) mainly distributed in Japan, China, Indonesia, and the Pacific Islands, the most chromosome number of which are $2n = 38$ and a few are $2n = 57$. *M. sacchriflorus* belonged to the representative species of *Eumiccanthus* Honda, whose utilization in hybridization breeding had been reported. Grassl (1967) demonstrated that the consanguinity of *M. floridus* had been introduced into *S. spontaneum and S. robustum* in New Guinea and Pacific Islands. In 1948, Li Xianwen obtained some hybrids with high biomass or high fiber content by crossing between *M. floridulus* and sugarcane parent POJ2725 and found three chromosome transmission methods are $2n +n$, $n+n$, and $n+2n$. In 1997, Burner reported that the chromosome transmission way in the crossing process is $n+2n$ between Sugarcane and *M. sinensis*. Only one way of $n+n$ was found in the crossing process between Sugarcane and *M. japonicus* or the progenies of F_1, F_2, and BC_2 of *M. japonicus*. The high resistance to dew or smut or cold of *M. sinensis* has been reported in some studies, so this species is considered of important breeding value. The Guangxi sugarcane research institute had four hybrids using *M. floridulus* to cross with sugarcane variety CP72-1210 (Huang et al., 1997). The F_1 progenies from the combination CP72-1210×*M. floridulus* have some excellent characteristic such as multiple tillers, vigorous growth, strong stress resistance, strong ratoon, high fiber content, high hardness, pipe, pith, and less juice, and the value of some main traits such as stem diameter and brix are between the parents; however, plant height and internode length performed over-parent heterosis, especially for plant height, but their sterile pollen hindered the further backcross utilization. Kariyasua Ohwi is widely distributed in Japan's mountainous areas; the chromosome number is $2n = 38$ generally and a few $2n = 114$. Adati et al. (1962) proposed that *Kariyasua* Ohwi could be a derivative of *Eumicacanthus* adapted to high altitude to contain many genes related to resistance to cold stress valuable for sugarcane breeding. Diandra Keng is mainly distributed in Sichuan, Yunnan, Tibet, northern India, and Nepal; the chromosome number is $2n = 40$. According to Grassl's studies (1972), *Diandra* Keng was suggested to assign to other genera. Then Daniels et al. (1975) suggested that it should be incorporated into the phylogenetic process of *Saccharum*, so it should be regarded as a kind of sugarcane.

2.3.3 The Introduction of *Sclerostachya* and Tts Breeding Value Analysis

Sclerostachya (Roxb) A. Camus is an intermediate type of *Saccharum* complex with hollow stems of 3 m in length. There are three main species, *Sclerostachya fusca* (Roxb) A. Camus, *Sclerostachya milroyi* bor. and *Sclerostachya ridleyi* (hack.) A. Camus was contained in this genus. Of which *S. fusca* was considered a representative species. About four common chromosome types were reported in the genus, $2n = 30, 34, 48$, and 98.

These genus species are distributed in northern India; The highest distribution altitude reaches 1,300 m in the Himalayas. For *S. fusca*, it is mainly distributed in India and Indochina Peninsula, and *S. ridleyi* was found in the Malay Peninsula. The main characteristics of *Sclerostachya* are hollow stems and spikelets arranged in pairs with pedicels, which are different from the spikelets arranged in pairs of

Saccharum and *Erianthus,* with only one having pedicels. In India, The progenies with a chromosome number of 55, 70, and 90 were obtained by crossing between *S. officinarum* and *Sclerostachya fusca*, whose chromosome transmission was considered as $n+n$, $n + 2n$ and $2n + n$ (Peng, 1990; Wen et al., 2001; Yu et al., 2004; Zhang et al., 2006).

2.3.4 The Introduction of *Narenga* and Its Breeding Value Analysis

Narenga bor is a perennial plant with tall, erect stem and lanceolate leaves often degenerated in the terminal. It is widely distributed in tropical areas of southeast Asia and northern India. *Narenga* has two distinct features: a bearded node and a dark yellow to the dark brown inflorescence. *Narenga* belongs to a small genus closely related to *Sclerostachya*, including *Narenga porphyrocoma* (Hance ex trimen) bor and *Narenga fallax* (BAL.) bor. Still, only the breeding characteristics of *Narenga porphyrocoma* were reported.

N. porphyrocoma, also known as "Caoxiemi", is native to Asia. It is mainly distributed in barren red soil area in the south of the Changjiang River in China, and the individuals can grow up to 2.5 m in some regions, but only grow up to 1.0 m in other regions. Its stem is thin and hollow, especially in the base. The length of its internode can reach 16 cm with a thicker upper part and no root point. Its long triangle bud is covered by thick scale, germinating of bud is difficult instead of underground stem propagation. Its leaf has a sharp blade edge and no developed midrib with 100−120 cm long and 0.7−1.2 cm wide, and the leave base is small, but the upper part of leaf is large, and the leaf sheath is tough to fall off because it is tightly wrapped around the stem. According to the previous reports, this species has low sugar content; for example, Yunnan 82-3 and Yingdehe only has 1.41% and 1.99% sugar content, respectively. This species has some excellent characters, such as barren tolerance, drought tolerance, rough growth, early maturity, strong tillering ability, erect lodging resistance, red rot resistance, smut resistance, mosaic disease resistance, and so on. This species' inflorescence belongs to panicle, which is very compact, lilac, 22–25 cm long, its pollen is well developed and with high quantity, and showed self-incompatibility. Still, heterotypic hybridization can produce a large number of seeds (Peng, 1990). Their leaves' upper surface is covered with villi for some populations, and the inflorescence rachis is surrounded by hairy. *N. porphyrocoma* has a small chromosome number of $2n=30$ (Peng, 1990; Wen et al., 2001). The National Germplasm Repository of Sugarcane, Some clones of *N. porphyrocoma* began to blossom in June and still had flower spikes until December. However, most of the clones bloom from October to November. *N. porphyrocoma* is easy to cross with these species belonging to *Saccharum*, such as *S. officinarum*, *S. spontaneum*, and *S. robustum*. Among them, the chromosome transmitting way is $n+n$ for the cross between *S. officinarum*×*N. porphyrocoma*.

Using *N. porphyrocoma* as male to cross with sugarcane varieties (CCT57-416, ROC1) had got a small number of hybrid progenies in the breeding plans implemented by Guangxi Sugarcane Research Institute (Huang et al., 1997). However, based on the number of seedlings obtained, *N. porphyrocoma* had higher compatibility with these species from *Saccharum* than *M. floridulus* and *E.rockii*. Its hybrid progenies showed more fabulous segregation characters and rich variation types. The progenies from sugarcane×*N. porphyrocoma* has a tall, erect stem with light green long smooth scattered leaves, leaf sheath covered by 57 hair, cylindrical internodes, protruding round buds connected with a flat-leaf scar under it the growth belt on the top. Under the same cultivation conditions, most progenies perform more tillers, vigorous growth, disease resistance, drought tolerance, barren tolerance, ratooning ability, etc. So breeders can select more multiple resistance and high biomass hybrids from their progenies. These hybrids' main traits, such as stem diameter and brix, were between both parents; transgressive phenomena were only found in plant height and internode length. However, some defects like high fiber content, high hardness, hollow, pith, less juice, low sugar, and so on must be improved by further crossing with *S.*

officinarum or elite sugarcane varieties; finally, some excellent hybrid progenies can be obtained by these ways above. But the serious problem is that the pollen sterility of F_1 progenies between *N. porphyrocoma* and Sugarcane influenced the further utilization of these elite progenies in sugarcane breeding plans.

References

AN R D, CHU L B, SUN Y F, et al., 2007. Breeding of a new sugarcane variety Yunzhe 99-155[J]. Sugarcane and Canesugar (3): 7-15.

BESSE P, MCINTYRE C L, 1999. Chromosome in situ hybridisation of ribosomal DNA in *Erianthus* sect. Ripidium species with varying chromosome numbers confirms x=10 in *Erianthus* sect. Ripidium[J]. Genome, 42 (2):270-273.

CHEN R K, XU L P, LIN Y Q, et al., 2011. Modern sugarcane genetic breeding[M]. Beijing: China Agricultural Press (in Chinese).

CHEN S F, HE J, ZHOU P H, et al., 2008. The Karyorypes of M*icanthus sinensis* and *M. floridulus*[J]. Acta Agriculturae Universitis Jiangxiensis (2):123-126.

CHEN X W, DENG H H, CHEN Y S, et al., 2000. Utilization of Badila in the breeding of YC-Series parents and new varieties of Sugarcane[J]. Sugarcane and Canesugar (6):1-5,43.

CHEN Y S, 2006. The Taiwan wild cane and its utilization in sugarcane breeding[J]. Sugar Crops of China (4):46-50.

FANGUY H P, 1974. CP65-357, a new sugarcane variety for Louisiana[J]. Proc. ASSCT (3):53-55.

GRASSL C O, 1967. Introgression between *Saccharum* and *Miscathus* in New Guinea and the Pacific area[J]. Proc. ISSCT (12):995-1003.

HUANG J Y, LIAO J X, ZHU G Y, 1997. Intergeneric copulatality of *Saccharum* L. with *Narenga porphyrocoma*, *Miscanthus floridulus* and *Erianthus kockii*, the morphology and isozyme analysis of their hybrid F_1 clones[J]. Southwest China

Journal of Agricultural Sciences, 10 (3):92-96.

JESWIET J, 1927. The history of sugarcane selection work in Java[J]. Proc. ISSCT (2):115-122.

LU Y H, D'HONT A, PAULET F, et al., 1994. Molecular diversity and genome structure in modern sugarcane varieties[J]. Euphytica, 78 (3):217-226.

LUO J S, 1984. Sugarcane science[C]. Guangzhou: Guangdong sugarcane Society.

PAN Y B, BURNER D M, WEI Q, et al., 2004. New Saccharum hybrid in *S. spontaneum* cytoplasm develops through a combination of conventional and molecular breeding approaches[J]. Plant Genetic Resources, 2 (2):131-139.

PENG S G, 1990. Sugarcane breeding[M]. Beijing: Agricultural Press (in Chinese).

PRICE S, 1967. Interspecific hybridization in sugarcane breeding[J]. Proc. ISSCT, (12):1021-1026.

WANG Q Z, 1979. Sugarcane Farming[M]. Taibei: National editorial Gallery.

WEN Y, 1998. Intergeneric crossing between *Saccharum* and related Plants and their chromosome behavior[J]. Sugarcane and Canesugar (3):1-7, 17.

WU C W, ZHAO P F, XIA H M, et al., 2014. Modern cross breeding and selection techniques in sugarcane[M]. Beijing: Science Press (in Chinese).

XIAO F H, LI F S, 1996. A study on wild relative species of Sugarcane (*Erianthus fulvus*)[J]. Sugarcane (Fujian), 3 (2):1-6.

XIAO F H, YANG Q H, ZHOU C P, et al., 1992. Observation on the Chromosomes and Botanical Characteristics of *Saccharum aruneinaceum*[J]. Journal of Yunnan Agricultural University (1): 59-62.

YANG Q H, HE S C, 1996. Chromosome number and geographical distribution of Saccharum spontaneum in Yunnan Province[J]. Sugarcane (1):10-13.

ZHANG M Q, WANG H Z, BAI C, et al., 2006. Genetic improvement and efficient breeding of sugar crops[M]. Beijing: China Agricultural Press (in Chinese).

3 The Hybrid Mode and Effect of Sugarcane Wild Resources

In 1887, Sottwedel in Java and Harrison and Bovell in Barbados saw that the seeds produced by sugarcane could sprout into seedlings; this discovery opened the prelude to the history of sexual cross-breeding of sugarcane (Luo, 1984). Since then, hybridization for sugarcane genetic improvement has been adopted in all sugarcane-producing countries. In the hybridization, *S. officinarum*, *S. barberi*, *S. sinense*, *S. spontaneum,* and *S. robustum* initially involved germplasm resources. The utilization methods include internal hybridization and interspecific hybridization, especially interspecific hybridization, which creates many excellent parents and varieties and lays the foundation for sugarcane breeding in the world. The hybridization between *S. officinarum* and *S. spontaneum* in Java created a series of excellent varieties with large stems, high sugar content, high yield, and disease resistance after three times of " Nobilization". Among all their progenies, the sugarcane cultivar POJ2878 was the most famous. In India, excellent varieties such as Co213, Co281, and Co290 were developed from *S. officinarum*, *S. barberi,* and *S. spontaneum*. In Hawaii, the USA, sugarcane cultivar H32-8560 and H49-5 were developed from the hybridization among *S. officinarum*, *S. spontaneum*, *S. barberi*, *S. robustum,* and *S. sinense*. Many new sugarcane varieties with high yield, excellent quality, and strong resistance were developed globally, which promoted the development of the cane sugar industry in the world.

Since the establishment of the Sugarcane Breeding Station in Hainan (HSBS) in 1953, sugarcane research organizations of China have successfully developed a series of new sugarcane varieties. More than 300 sugarcane varieties have been developed, which has promoted the sugar industry development in China. For example, cultivars CYT 57-423 and CYT 93-159 in Guangdong, CGT11 and CGT15 in Guangxi, CMT 69-421 in Fujian, CYZ 71-388 and CYZ 89-151 in Yunnan, CCZ

13 in Sichuan, and so on. From 1998 to 1999, the five varieties with the largest planting area in China were CGT11, ROC10, Xuan 3, CYT 63-237, and ROC16, accounting for 21.04%, 9.27%, 8.72%, 6.77%, and 6.53% of the total area of China respectively.

Hybridization is needed to develop a new cultivar; however, the propagation is a vegetative process, so each plant within a cultivar should be genetically consistent unless there is a mutation. Most of the varieties are derived from 3-5 sugarcane original *S. officinarum* and then developed by cross and backcross. Their genetic background is similar, which leads to the inbreeding of sugarcane varieties, narrow genetic basis, and close genetic relationships. Consequently, it is challenging to gain breakthroughs in comprehensive traits such as yield, sugar content, and resistance in sugarcane breeding in the recent 50 years. Therefore, sugarcane breeding organizations worldwide pay great attention to collecting, research, and utilizing sugarcane germplasm resources to expand blood relationships, enrich genetic basis, and create breakthrough parent materials and excellent varieties. Wu (2005) found that sugarcane germplasm resources had different cross-utilization methods and different offspring performances. And put forward the views of parent improvement, innovation, and independent parent system. He believed that through new interspecific hybridization within the genus of sugarcane, the new generation F_1 continues to equal cross. A new independent parent system can be cultivated through two or three times equal crossing. The successful breeding of a new independent parent system can lay an essential foundation for breeding breakthrough sugarcane varieties.

3.1 Ways and Effects of Hybrid Utilization of Wild Species of Related Genus of Sugarcane

Commercial characters of modern sugarcane cultivars come from *S. officinarum*, while adaptability and stress tolerance come from wild species. In the hundred years of sugarcane genetic improvement, the exploration and utilization of excellent germplasm resources contributed significantly. Sexual hybridization is used to create new germplasm and breed new sugarcane varieties all over the world. For example, POJ2878, Nco310, Co419, and ROC22 are outstanding interspecific hybridization and noble breeding in the first half of the 20th century. They were developed from 2, 3, or 5 germplasm within *S. officinarum, S. barberi, S. sinense, S. spontaneum,* and *S. robustum*. However, in recent 50 years, there was no breakthrough in sugarcane breeding. Close genetic relationships and lack of germplasm are possible reasons. Therefore, in the past 20 years, various sugarcane-producing countries have successfully carried out the cross-utilization of many related wild species to create the glory of sugarcane breeding again.

Wild germplasm of related genera of sugarcane mainly includes *Miscanthus Sinensis, Erianthus rufipilus, Narenga porphyrocoma,* and *Sclerostachya (Roxb)* A. Camus. Among them, *Miscanthus Sinensis* is the most widely used, and there are few reports on the hybrid utilization of *Sclerostachya (Roxb)* A. Camus.

3.1.1 Hybridization of *Erianthus Michx.* Sect. *Ripidium Henrard* and Their Achievements

About 20 *Erianthus rufipilus* are distributed in the temperate zone, subtropical zone, and tropical zone. There are 4 species in China: *E. formosanus, E. rockii, E. fulvus* and *E. arundinaceum*. Many reports on the utilization of *E. fulvus, E. arundinaceum,* and *E. rockii*, but few reports on the others.

3.1.1.1 The hybrid utilization mode and effect of *Erianthus rufipilus*

Erianthus rufipilus ($2n = 20$) is a wild species in *Erianthus*. There is no variation among clones and having the least number of chromosomes in the sugarcane complex. It is very beneficial to explore the genetic of chromosomes and make the molecular marker-assisted selection in hybridization and utilization. *Erianthus rufipilus* distributes in unique geographical areas and has the characters of wide adaptability, drought resistance, cold resistance, barren resistance, and high brix, which is superior to other sugarcane-related wild species (Li et al., 2003), and is of great potential in sugarcane genetic improvement.

Utilization of *Erianthus rufipilus* (Table 3-1) included two approaches, i.e., sugarcane complex × *Erianthus rufipilus* and *S. spontaneum* × *Erianthus rufipilus*. Among them, there is more report on sugarcane complex × *Erianthus rufipilus*. As early as 1927, Rankle crossed the offspring EK28 (POJ100 ×EK2) of *S. officinarum* to *Erianthus rufipilus* and obtained offsprings. In 2001, Li Fushang crossed commercial cultivars (CYC 89-9, Lianghe78-121(CLH7-121), CP72-1210, CHN56-12, ROC16, and ROC20) to Kunming *Erianthus rufipilus*, and made significant progress, and obtained a batch of innovative materials with excellent yield, sugar content, and cold resistance to adapt to high altitude areas. Among them, "CYC 89-9 × Kunming *Erianthus rufipilus* " was an excellent cross combination, which generated 132 seedlings, and 47 of them were identified as true hybrids. These progenies endured low temperatures in winter and survived, indicating that they have strong cold tolerance. Simultaneously, some of them also have excellent characteristics such as drought resistance, high cane yield, and high sugar content, which are expected to be used as new specific germplasm for genetic improvement for stress resistance and wide range adaption. The progeny Yunnong 01-58 was advanced to the tenth regional test of sugarcane varieties in Yunnan, and its mean cane yield across plant and ratoon crops reached 77.99 t/hm^2 and was 22% lower than that of the control. Its mean sugar content was 14.72%, 0.39−0.5 percentage points lower than that of control, but it showed drought tolerance and cold tolerance at high altitude areas. It was finally released for high-altitude cane areas. There is only one report on the cross-utilization of *S. spontaneum* × *Erianthus rufipilus*. In 1953, Lagerfan crossed *Erianthus rufipilus* with *S. spontaneum* and obtained offspring with $2n + n$ chromosome number (Zhang et al., 2006). The above research showed that crossing *Erianthus rufipilus* as a male parent to wild species and hybrids in sugarcane could obtain progenies. Sugarcane complex × Kunming *Erianthus rufipilus* was a successful approach that a range of cross combinations was used for producing fuzzy, and a cultivar with outstanding cold tolerance was released.

Table 3-1 Cross utilization statistics of *Erianthus rufipilus*

Cross combination	Hybridization type	Effect
EK28 × *Erianthus rufipilus*	Saccharum complex × *Erianthus rufipilus*	Obtained hybrid offspring
CYC 89-9 × Kunming *Erianthus rufipilus*	Saccharum complex × *Erianthus rufipilus*	Excellent performance, breeding a variety (Yunnong 01-58)
CLH 78-121 × Kunming *Erianthus rufipilus*	Saccharum complex × *Erianthus rufipilus*	Obtained hybrid offspring
CP72-1210 × Kunming *Erianthus rufipilus*	Saccharum complex × *Erianthus rufipilus*	Obtained hybrid offspring
CHN56-12 × Kunming *Erianthus rufipilus*	Saccharum complex × *Erianthus rufipilus*	Obtained hybrid offspring
ROC16 × Kunming *Erianthus rufipilus*	Saccharum complex × *Erianthus rufipilus*	Obtained hybrid offspring
ROC20 × Kunming *Erianthus rufipilus*	Saccharum complex × *Erianthus rufipilus*	Obtained hybrid offspring
S. officinarum × *Erianthus rufipilus*	*S. officinarum* × *Erianthus rufipilus*	Obtained hybrid offspring
S. spontaneum × *Erianthus rufipilus*	*S. spontaneum* × *Erianthus rufipilus*	Obtained hybrid offspring

Source: Wu et al., 2014.

3.1.1.2 Utilization of *Erianthus rockii* and its achievements

E. rockii is a unique and precious wild germplasm of related genus in China, which is only distributed in Sichuan, Yunnan, and Tibet of China and grows in the dry hillside grassland altitude 50–2,700 m. It has excellent characteristics such as drought tolerance, cold tolerance, barren adaption, good ratooning ability, and rust resistance. It shows stronger resistance to diseases and pests and better ratooning ability than the *S. spontaneum* (Wang et al., 2008). It is an essential resource for the genetic improvement for disease resistance, particularly for rust resistance. There are three reported hybrid utilization modes of *E. rockii* (Table 3-2, Table 3-3),

Saccharum complex × *E. rockii*, *S. officinarum* × *E. rockii*, and landrace × *E. rockii*, but there are few further reports.

(1) Saccharum complex × *Erianthus rockii*

Huang et al. (1997) tried the distant hybridization between sugarcane cultivars and *E. rockii* and successfully obtained fuzzy from two combinations CP72-1210× *Erianthus rockii* 86-10, and ROC1× *Erianthus rockii* 86-10 and obtained six seedlings. The performance of stalk diameter and brix of the hybrids were within the performance of their parents. The stalk diameter of the hybrids was small, the stem hardness was moderate, the leaves were short, smooth, and wide, and there was no hair on the leaf sheath, the bud was round and small, and the bud wing was large. After that, the hybrids were backcrossed to *Erianthus rockii* again and obtained three seedlings. In this backcross generation, the stalk diameter became smaller with a pink ribbon and hollow in the stalk, their leaves were short and wide, and in the color of light blue, their brix was in the range of 7.5%–16.5%. These were true hybrids according to morphology and isoenzyme identification. Ruili Breeding Station of Yunnan Sugarcane Research Institute of Yunnan Academy of Agricultural Sciences (Tao, 1996) crossed sugarcane cultivar CGT11 to *Erianthus rocki*i and obtained 14 seedlings. One of them, CYR91-2008, was advanced as a promising clone. And further, CYR91-2008 was crossed to CYR93-3148 (offspring of *S. spontaneum* and *Erianthus arundinaceus*) and developed CYR99-48, an offspring that contained genetic background from *S. spontaneum* and *Erianthus arundinaceus* and *Erianthus rockii*. It showed excellent resistance to sugarcane smut and leaf spot (Yang, 2004). CYR 99-48 was further crossed to sugarcane cultivar F172 and obtained two clones, CYR 00-7 and CYR 00-19. Zhu et al. (1996) crossed excellent sugarcane cultivar to *Erianthus rockii*, obtained fuzzy from 2 cross combinations and five seedlings. They were true hybrids, and their performance was similar to that reported by Huang (1997).

Table 3-2 Status of *Saccharum* complex × *Erianthus rockii*

Offspring	Cross combination	Hybridization type	Note
CGTF191-1	CP72-1210×*Erianthus rockii* 86-10	Saccharum complex×*E.rockii*	Guangxi, China
CGTF191-12	ROC1×*Erianthus rockii* 86-10	Saccharum complex×*E.rockii*	Guangxi, China
CYR 91-2008	CGT11×*Erianthus rockii*	Saccharum complex×*E.rockii*	Ruili, China
CYR 00-7, CYR 00-19	F172×*Erianthus rockii*	Saccharum complex×*E.rockii*	Ruili, China

Source: Wu et al., 2014.

(2) S.officinarum× Erianthus rockii

Wang et al. (2008) reported that 9 *S. officinarum* and 4 *Erianthus rockii* were crossed and obtained fuzzy from 10 combinations (Table 3-3), and the seedlings were obtained, of which the maximum number of seedlings was 178 seedlings/tassel, and the minimum number was only one seedling/tassel. A batch of F_1 hybrids was obtained after selection. Due to pollen abortion of F_1 progenies, some excellent BC_1 clones with strong drought resistance and high rust resistance were obtained by backcrossing F_1 clones (female). The number of seedlings per tassel of backcrossing offspring was more than that of F_1 progenies, and the selection rate was higher. The agronomic characteristics, flowering characteristics, selection of hybrid offspring, and backcrossing offspring's performance were reported in detail.

(3) Landrace × *Erianthus rockii*

Wang et al. (2008) crossed 9 *S. officinarum* to 4 *Erianthus rockii*, and at the same time also designed a cross combination that crossing between Taoshan chewing cane and Yunnan *Erianthus rockii* 83-224 (Table 3-3), and obtained four seedlings, and selected one clone which was identified by authenticity as the progeny contained genetic background of *Erianthus rockii* but not the true hybrid of Yunnan *Erianthus rockii* 83-224.

Table 3-3 Statistics of basic cross-utilization of *Erianthus rockii* in Yunnan

Cross combination	Hybridization type	Note
Luohancane × *Erianthus rockii* /Yunnan 95-19	*S. officinarum* × *E. rockii*	Kaiyuan, China
Niucane × *Erianthus rockii* /Yunnan 95-20	*S. officinarum* ×*E. rockii*	Kaiyuan, China
Fijian × *Erianthus rockii* /Yunnan 83-224	*S. officinarum* ×*E. rockii*	Kaiyuan, China
48Mouna × *Erianthus rockii* /Yunnan 95-19	*S. officinarum* ×*E. rockii*	Kaiyuan, China
NC32 × *Erianthus rockii* /Yunnan 83-224	*S. officinarum* ×*E. rockii*	Kaiyuan, China
51NG90 × *Erianthus rockii* /Sichuan 92-40	*S. officinarum* ×*E. rockii*	Kaiyuan, China
28NG16 × *Erianthus rockii* /Yunnan 83-224	*S. officinarum* ×*E. rockii*	Kaiyuan, China
Barviwlpt × *Erianthus rockii* /Yunnan 82-224	*S. officinarum* ×*E. rockii*	Kaiyuan, China
96NG16 × *Erianthus rockii* /Yunnan 83-224	*S. officinarum* ×*E. rockii*	Kaiyuan, China
96NG16 × *Erianthus rockii* /Yunnan 95-20	*S. officinarum* ×*E. rockii*	Kaiyuan, China
Taoshan chewing cane × *Erianthus rockii* /Yunnan 83-224	Landrace ×*E. rockii*	Kaiyuan, China

Source: Wu et al., 2014.

The results showed that, although the pollen of *Erianthus rockii* is excellent and abundant, it is easy to cross with *S. officinarum*, landrace, or hybrids in sugarcane as a male parent. Still, because it is different from sugarcane, the hybridization rate is not high. The compatibility between *S. officinarum* and *Erianthus rockii* is poor. The seed setting rate of their hybridization is relatively low, so many hybridization combinations are needed to obtain enough offspring. *Erianthus rockii* are short, with thin stalk and low brix, but drought resistance and rust resistance are better in the offspring. The BC_1 true hybrids obtained by backcross between F_1 generation and sugarcane varieties are superior to *Erianthus rockii* in plant height, stalk diameter and brix. Their unhealthy features are easy to be improved. Its BC_1 shows strong drought resistance, high rust resistance, and high yield potential.

3.1.1.3 The Way and Effect of hybrid utilization of *Erianthus arundinaceum*

Erianthus arundinaceus is a critical wild resource in the sugarcane complex, which has excellent characteristics such as good shoots form, strong germination, many tillers, strong vigor, good ratooning, wide adaptability, strong stress tolerance, muscular disease, and insect resistance, etc. As early as 1885, Scotwedel tried to cross sugarcane with *Erianthus arundinaceus* but failed. Some countries that started sugarcane breeding research earlier, such as Barbados, the United States, South Africa, etc., have researched hybrid utilization of *Erianthus arundinaceus*. Still, no other hybrid or backcross utilization has been carried out for developing sugarcane cultivar. In recent years, Hainan Sugarcane Breeding Station of Guangzhou Sugarcane Industry Research Institute (HSBS) and Sugarcane Research Institute of Yunnan Academy of Agricultural Sciences (YSRI) have made significant progress by using *Erianthus arundinaceus* and have successively developed some sugarcane clones with the genetic background of *Erianthus arundinaceus* (Liu, 1992; Liu et al., 2007). Although *Erianthus arundinaceus* has a large amount of pollen and good pollen fertility, due to its distant genetic relationship with sugarcane, the seed setting rate of hybridization is meager. Even if some hybrid offspring can be obtained, the pollen fertility is lacking or even sterile (Xiao et al., 1995), which brings great difficulties to the utilization of *Erianthus arundinaceus*. At present, only the early generations of *Erianthus arundinaceus* (F_1, BC_1, or BC_2) have been obtained (Wang et al., 2007; Liu et al., 2007), but there is no report indicating developing sugarcane varieties by using the *Erianthus arundinaceus*. Therefore, utilization is still a hot topic in sugarcane breeding. Till now, there have been three reported hybrid utilization modes of *Erianthus arundinaceus*.

(1) Saccharum complex × *Erianthus arundinaceus*

Erianthus arundinaceus is the most widely used wild species of sugarcane-related genera. There are many reports on the hybrid utilization of wild germplasm of sugarcane-related genera in China (Li, 2010), and many progenies have been developed (Table 3-4). Hainan Sugarcane Breeding Station (HSBS) began to cross sugarcane hybrids with *Erianthus arundinaceus* in 1956. By the end of the 20th century, CYC 57-25 and CYC 80-83 were developed successively. Some excellent sugarcane varieties, such as Ganzhe 1(CGZ 1) and CGZ 8 were developed by crossing or backcrossing the sugarcane cultivar with CYC 57-25. However, later chromosome identification results showed that CYC 57-25 was not a true hybrid of *Erianthus arundinaceus* (Wu et al., 1987), so the cultivars could not be the offspring *Erianthus arundinaceus* (Fu et al., 2003). Ruili Sugarcane Breeding Station (RSBS) developed the hybrid utilization of *Erianthus arundinaceus* late but used more *Erianthus arundinaceus*. To make full use of the conditions of Inland Sugarcane Hybrid Breeding Bases, the Ruili Sugarcane Breeding Station of Yunnan Academy of Agricultural Sciences (RSBS) has made many explorations in the hybrid utilization of hybrids and *Erianthus arundinaceus* since the 1980s. Different hybrids were used to hybridize with different *Erianthus arundinaceus*, such as Manhao *Erianthus arundinaceus*, Funing *Erianthus arundinaceus Erianthus arundinaceus* 180, a batch of hybrid offspring such as CYR 80-1180 were created. Although sugarcane varieties have not been developed yet, it lays a foundation for further exploring the excellent genes of *Erianthus arundinaceus*. In 2002, Yunnan Sugarcane Research Institute of Yunnan Academy of Agricultural Sciences (YSRI) successfully obtained seedlings by crossing two cultivars with three clones of *Erianthus arundinaceus*, and the number of seedlings was significantly more than that of *S. officinarum* × *Erianthus arundinaceus* (Wang et al., 2007).

Table 3-4 Utilization and Effect of hybrid × *Erianthus arundinaceus*

Cross combination	Hybrid site	Progeny
CP34-120 × *Erianthus arundinaceus*	Yacheng, Hainan	CYC 57-25
S17 × *Erianthus arundinaceus*	Yacheng, Hainan	CYC 72-399
CYT 57-423 × *Erianthus arundinaceus*	Yacheng, Hainan	CYC 73-512
Co1001 × *Erianthus arundinaceus* of Yacheng	Yacheng, Hainan	CYC 90-31
CP72-1210 × *Erianthus arundinaceus*	Yacheng, Hainan	CYC 90-4, CYC 90-11

		continued
Cross combination	Hybrid site	Progeny
CYT 54-18 ×*Erianthus arundinaceus* in Manghao of Yunnan	Ruili, Yunnan	CYR 80-114
CYR 80-15 × *Erianthus arundinaceus* in Funing of Yunnan	Ruili, Yunnan	CYR 93-3148
CYR 00-7 × *Erianthus arundinaceus* 180	Ruili, Yunnan	CYR 06-75
Co285 × Jiangxi *Erianthus arundinaceus* 79-02	Kaiyuan, Yunnan	115 seedlings are all false hybrids
Co285 × Guangxi *Erianthus arundinaceus* 84-16	Kaiyuan, Yunnan	45 seedlings were all false hybrids
Co419 × Yunnan *Erianthus arundinaceus* 83-158	Kaiyuan, Yunnan	26 seedlings, 57% of them are true hybrids

Source: Wu et al., 2014.

(2) *S.officinarum* × *Erianthus arundinaceus*

There are many cross combinations under the type of *S. officinarum* × *Erianthus arundinaceus*, resulting in a large number of offspring seedlings and clones (Table 3-5). In 1941, Janach Anmuru used *S. officinarum* to cross with *Erianthnus arundinaceus*, and analyzed the chromosome of the obtained progenies. The genetic pattern was $n + n$. In Hainan Sugarcane Breeding Station of China (HSBS), 5 *S. officinarum* were crossed with ten different *Erianthus arundinaceus*, and 11 cross combinations including 50Uahiapele × *Erianthus arundinaceus*, Badila × *Erianthus arundinaceus*, Akoki22 × *Erianthus arundinaceus,* and 27MQ1124 × *Erianthus arundinaceus,* etc., generated fuzzy (Liu et al., 2007), and some of the progenies were continued to be crossed with commercial varieties, and the true hybrids of F_1 and F_2 generations verified by molecular marker technology. Among them, F_1 generations included CYC 96-35, CYC 96-36, CYC 96-60, CYC 95-41, CYC 96-40, CYC 96-55, CYC 96-66, CYC 96-45, and CYC 96-69, etc., and F_2 generation included CYC 01-3. A total of 11 *S. officinarum* (Crystalina, Muckche, 48Mouna, Fiji, Keong Java, NC32, Manjar, 96NG16, Badila, YuenanNiucane, and Luohancane) have been used in Yunnan Sugarcane Research Institute of Yunnan Academy of

Agricultural Sciences (YSRI) since 1998 to cross with different types of *Erianthus arundinaceus* collected from Yunnan, Hainan, Guangdong, Jiangxi, and Fujian province and seedlings were obtained. Among these, a batch of germplasm with freezing tolerance and high smut resistance was developed. A batch of excellent innovative germplasm with F_1, BC_1, BC_2, BC_3, and BC_4 identified by molecular identification were obtained by backcrossing with commercial cultivars, which laid a solid foundation for the utilization of germplasm with the genetic background of *Erianthus arundinaceus* (Wang et al., 2003; 2007).

Table 3-5 Statistics of basic hybridization and utilization of *S. officinarum* × *Erianthus arundinaceus*

Cross combination	Hybrid site	Progeny
50Uahiapele × *Erianthus arundinaceus*	Yacheng, Hainan	CYC 96-35, CYC 96-36, CYC 96-60
Akoki22 × *Erianthus arundinaceus*	Yacheng, Hainan	CYC 96-32
Badila × *Erianthus arundinaceus*	Yacheng, Hainan	CYC 73-07
Badila × Hainan 92-77+ Hainan 92-79	Yacheng, Hainan	CYC 73-557
B.cheribon × *Erianthus arundinaceus* in Hainan	Yacheng, Hainan	CYC 73-906
Badila × Yacheng *Erianthus arundinaceus*	Yacheng, Hainan	CYC 75-283, CYC 75-285
Badila × Hainan 92-105 + Hainan 92-77 + Hainan 92-79	Yacheng, Hainan	CYC 95-40, CYC 95-41, CYC 95-42, CYC 95-43
Badila × *Erianthus arundinaceus* (Hainan 92-105)	Yacheng, Hainan	CYC 95-46
Badila × Hainan 92-77 + Hainan 92-79	Yacheng, Hainan	CYC 96-40, CYC 96-41
Badila × *Erianthus arundinaceus*	Yacheng, Hainan	CYC 96-45, CYC 96-55
Badila × *Erianthus arundinaceus* (Hainan 92-105)	Yacheng, Hainan	CYC 96-65, CYC 96-66, CYC 96-67, CYC 96-68
Badila × *Erianthus arundinaceus* /Hainan 90-215	Yacheng, Hainan	CYC 96-66
Badila × *Erianthus arundinaceus* /(Hainan 92-79)	Yacheng, Hainan	CYC 96-69
Luohancane × *Erianthus arundinaceus*/Haiban 92-84	Kaiyuan, Yunnan	CYZ 00-117, CYZ 00-118

Source: Wu et al., 2014.

(3) *S.sinense* and landrace × *Erianthus arundinaceus*

To further explore the effect of hybrid utilization of *Erianthus arundinaceus*, the Hainan Sugarcane Breeding Station (HSBS) has continuously expanded the scope of hybrid utilization (Table 3-6). Since the 1990s, a series of *S. sinense* and landrace have been hybridized with *Erianthus arundinaceus*, including Songxi Bainiancane, Guangxi zhucane, Guangdong zhucane and Jiangxi zhucane (Liu et al., 2000). A range of clones included CYC 95-21, CYC 95-26, CYC 95-27, CYC 95-30, CYC 95-31, CYC 95-34, CYC 95-35, CYC 96-61, and CYC 96-62 were developed from the hybridization.

Table 3-6 Statistics of basic hybridization of *S. sinense* and landrace × *Erianthus arundinaceus*

Cross combination	Hybridization mode	Progenies
Jiangxi zhucane × *Erianthus arundinaceus*	*S. sinense* × *Erianthus arundinaceus*	CYC 95-21, CYC 96-61, CYC 96-62
Guangxi zhucane × *Erianthus arundinaceus*	*S. sinense* × *Erianthus arundinaceus*	CYC 95-30, CYC 95-31
Bainiancane × *Erianthus arundinaceus*	Landrace × *Erianthus arundinaceus*	CYC 95-26, CYC 95-27
Guangdong zhucane × *Erianthus arundinaceus*	*S. sinense* × *Erianthus arundinaceus*	CYC 95-34, CYC 95-35
Songxi Bainiancane × Sichuan 79-I-9, etc	*S. sinense* × *Erianthus arundinaceus*	There were 8 seedlings, and 2 were advanced
Songxi Bainiancane × Yacheng *Erianthus arundinaceus* No.1 and so on	*S. sinense* × *Erianthus arundinaceus*	There were 4 seedlings, and 2 were advanced
Guangxi zhucane × Sichuan 79-I-11, etc	*S. sinense* × *Erianthus arundinaceus*	Eleven seedlings were obtained, and two of them were advanced
Guangdong zhucane × *Erianthus arundinaceus*	*S. sinense* × *Erianthus arundinaceus*	There were 9 seedlings, and 2 were advanced
Jiangxi zhucane × *Erianthus arundinaceus*	*S. sinense* × *Erianthus arundinaceus*	There were 8 seedlings, and 2 were advanced
Jiangxi zhucane × Yacheng *Erianthus arundinaceus* No.2 and so on	*S. sinense* × *Erianthus arundinaceus*	Seeds did not emerge

Source: Wu et al., 2014.

(4) *S.robustum* × *Erianthus arundinaceus*

There is only one reported case of wild species with *S. robustum* × *Erianthus arundinaceus*. In 2007, Hainan Sugarcane Breeding Station (HSBS) used wild species with *S. robustum* to hybridize with *Erianthus arundinaceus* and successfully obtained a fuzz of NG77-1 × Hainan 92-92 (Liu et al., 2007), and generated 3 seedlings. The clone CYC96-63 was one of the seedlings. Still, there has been no further report on its utilization.

The results of hybrid utilization of *Erianthus arundinaceus* showed that: When sugarcane (*S. officinarum*, landrace, wild species or hybrids) was hybridized with *Erianthus arundinaceus*, the amount of F_1 pollen was small, the fertility was poor, the hybridization compatibility was low, the seed setting rate was very low, and the number of seedlings was small; The BC_1 generation of *Erianthus arundinaceus* produced abundant pollen and was well-developed male parental clone, which basically eliminated the cross incompatibility with cultivated varieties, indicating that the possibility of breeding new sugarcane varieties by *Erianthus arundinaceus* was further increased; "*S. officinarum* × *Erianthus arundinaceus*" is more accessible to obtain seedlings and true hybrids of *Erianthus arundinaceus* than" commercial variety × *Erianthus arundinaceus* ", while commercial variety are more suitable as backcross parents; Judging from the agronomic characters of BC_1 and BC_2 of *Erianthus arundinaceus*, unfavorable characteristics of *Erianthus arundinaceus* such as low sugar and thin stalk tended to be improved, however, there are obvious unfavorable traits such as lateral buds, obvious air roots, difficult defoliation and vulnerable to pests, and most of them have the disadvantages of big pith and heavy hair on leaf sheath; From the main characters of BC_1 and BC_2 of *Erianthus arundinaceus*, the excellent characteristics of *Erianthus arundinaceus* with strong vigor can be inherited stably, and the unfavorable characteristics of low brix is easy to improve. Simultaneously, the pith of *Erianthus arundinaceus* reduced with the increase of the generation of backcrossing. The ultimate goal of hybrid utilization of sugarcane germplasm resources is to infiltrate the excellent characteristics of *Erianthus arundinaceus* into a modern cultivar. The creation of F_1, BC_1, and BC_2 is the beginning of their utilization in sugarcane genetic improvement. There is still a lot of work to be done in breeding new sugarcane varieties with the genetic background of *Erianthus arundinaceus*. Sugarcane breeders must intensify development and utilization to broaden the genetic basis of sugarcane cultivated original species for developing excellent modern cultivar.

3.1.2 The Way and Effect of Hybridization of *Miscanthus*

Miscanthus Anderss can be divided into *Trierrhena* Honda, *Eumiscanthus* Honda, *Kariyasua* Ohwi, and *Diandra* Keng. Among them, there are reports on the hybrid utilization of *Eumiscanthus* Honda. There are mainly two species of *Eumiscanthus* Honda, namely *M. floridulus* Warb. Ex Schum.et Laut and *M. sinensis* Anderss. *M. Floridulus* is the representative species. Grassl (1967) has confirmed that in New Guinea and Pacific Islands, the gene of *M. floridulus* has infiltrated into wild species of *S. spontaneum* and *S. robustum*. The hybrid utilization mode of *Miscanthus* Anderss is *Saccharum* complex × *Miscanthus sinensis* Anderss (Table 3-7).

Li (1948) in Taiwan Province of China crossed *M. floridulus Warb.* ex *Schum. et Laut* to sugarcane cultivar POJ2725 and the offspring obtained were improved in biomass and fiber content. Zhu and Huang of Guangxi Sugarcane Research Institute (1996) crossed the commercial sugarcane cultivar to *Miscanthus Sinensis* anderss and obtained five seedlings from two cross combinations. The results showed that the hybrid offspring showed scattered leaves, sharp leaf margins, more rust, bushy stalks, easy flowering, thin stalk, and the brix was in a range of 8.5%–16.0%. Huang et al. (1997) crossed three sugarcane hybrid varieties to *Miscanthus Sinensis Anderss*, but the seed setting and germination rates were low. In the hybridization, CGT87-86 and CFN87-2 were crossed as female parents. Each obtained one tassel, but the seeds did not germinate; Only CP72-1210× *Miscanthus Sinensis Anderss* had 26 seeds germinated, and four were obtained. F_1 showed tall plants, large nodes, cylindrical internodes, and more white stalks. Leaves scattered, thick and hairy; The leaf margin is sharp, the bud is small, round or oval, the lower leaf mark is flat, the upper leaf mark is lower than the growth loop, the stalk was cyan, soft in quality, with more juice and higher brix. Stalk diameter and brix are between parents, and plant height and internode length surpassed their parents. Because of the distant genetic relationship between *Miscanthu*s and sugarcane, the variation range of chromosome number is wide, $2n = 38-114$ (Wen, 1998). The pollen of hybrid offspring is sterile, which brings difficulties for further backcross utilization.

Table 3-7 Utilization of *Saccharum* complex × *Miscanthus sinensis*

Cross combination	Location	Progeny
CGT87-86 × *M. floridulus Warb. Ex schum. Et Laut*	Nanning, China	Seeds did not germinate
CFN 87-2 × *M. floridulus Warb. Ex schum. Et Laut*	Nanning, China	Seeds did not germinate
CP72-1210 × *M. floridulus Warb. Ex schum. Et Laut*	Nanning, China	CGT94-26
POJ2725 × *M. floridulus Warb. Ex schum. Et Laut*	Taiwan, China	Showed high biomass and high fiber content

Source: Wu et al., 2014.

3.1.3 The Way and Effect of Hybrid Utilization of *Narenga Porphyrocoma*

Narenga porphyrocoma has early maturity, thick tillering, erect, lodging resistance, resistance to red rot, smut, mosaic disease, and waterlogging resistance; these are desirable traits for modern sugarcane cultivar. There are two species: *Narenga Porphyromoma (hanceex trimen) Bor* and *Narenga Fallax (bal.) Bor*. There are few reports on the cross-utilization of *Narenga porphyrocoma*. There are three utilization modes: Sclerotinia× *Narenga porphyrocoma*, *S. officinarum*× *Narenga porphyrocoma,* and Saccharum complex × *Narenga porphyrocoma*. In 1952, Raghavan used *Sclerostachya* and *S.Vellai* to study *Sclerostachya* × *Narenga* and *S.Vellai* × *Narenga*, but no further report on the follow-up results. In the 1990s, Sugarcane Research Institute of Guangxi Academy of Agricultural Sciences (GXSRI) carried out the cross-study of Saccharum complex × *Narenga porphyrocoma* (Table

3-8), using sugarcane hybrid varieties CCT 57-416, ROC1, and CGT81-660 as female parents to cross with *Narenga porphyrocoma* respectively, among which CCT 57-416× *Narenga porphyrocoma*, 44 seedlings were obtained, and 9 were selected; ROC1× *Narenga porphyrocoma*, 9 seedlings were obtained, and 2 were selected; The hybrid seeds of CGT81-660× *Narenga porphyrocoma* did not germinate (Huang et al., 1997). Zhu et al. (1996) crossed sugarcane cultivar to *Narenga porphyrocoma*, obtained fuzzy from three cross combinations, 53 seedlings were obtained, and 11 clones were selected. The experimental results showed that the true hybrids of *Narenga porphyrocoma* showed higher plants, fewer tillers, greater stalk hardness, green cane, more sheath hairs, stalk diameter of 0.92–1.82 cm, and brix of 5.5%–11.5%.

The results showed that crossing *Narenga porphyrocoma* to sugarcane as a male parent, their offspring performed tall and erect plants; The leaves are scattered, light green, long, smooth and glabrous; heavy hair on leaf sheath; cylindrical internode; buds are in the shape of round and convex; Lower flat-leaf mark, upper flat growth zone. Because the chromosome $2n=30$ of *Narenga porphyrocoma* is relatively fixed (Wen, 1998), the cross-compatibility between *Narenga porphyrocoma* and sugarcane is higher *Miscanthus* with sugarcane, and the hybrid seeds have more germination, more offspring advanced (Huang et al., 1997).

Table 3-8 Utilization of hybrid × *Narenga porphyrocoma*

Cross combination	Location	Progeny
CCT 57-416 × *Narenga porphyrocoma* 89-13	Nanning, Guangxi	CGT94-15
ROC1 × *Narenga porphyrocoma* 89-13	Nanning, Guangxi	CGT94-25
CGT81-660 × *Narenga porphyrocoma* 89-13	Nanning, Guangxi	No seedlings

Source: Wu et al., 2014.

3.2 The Hybrid Utilization Mode and Effect of Wild Saccharum Species

Traditional classification methods divide wild species of *Saccharum* L. into the following three species: *S. spontaneum, S. robustum,* and *S. edule. S. spontaneum* and *S. robustum* are important wild germplasm resources in sugarcane, which have played a vital role in sugarcane genetic improvement. It can be said that the history of sexual cross-breeding of sugarcane is the history of cross utilization of wild germplasm resources. *S. edule* wild species with fleshy tassels are only used as food plants in New Guinea, and their breeding value has not been found yet.

3.2.1 The Utilization of *S. spontaneum*

S. spontaneum is the most critical wild germplasm resource in sugarcane genetic improvement, modern sugarcane contains 15%–20% genetic background of *S. spontaneum* (D'Hont et al, 1996; Piperidis et al., 2001). *S. spontaneum* is a complex polyploid plant, is one of the wild species with the most breeding and research value among sugarcane and its related genera, and plays a vital role in sugarcane breeding. Since the beginning of the 20th century, Indonesia (Java) and India have made the most significant progress in modern sugarcane improvement by first crossing and backcrossing with tropical sugarcane species, and developed the famous cultivars POJ2878 (BC_2 of Java *S. spontaneum*), Co290 (BC_3 of India *S. spontaneum*), Co419 (BC_3 of India and Java *S. spontaneum*) and CP49-50 (BC_3 of India *S. spontaneum*) POJ2878, Co290, Co419 and CP49-50 have not only become widely popularized sugarcane varieties in the world, but also have been used as outstanding parental clones in many sugarcane breeding institutions. According to our understanding, POJ2878, Co290, Co419, and CP49-50 are directly used as parents (including female and male parents, as many as 181, 29, 81, and 70 varieties have been developed, respectively, such as China, Brazil, India, Australia, Thailand, Cuba, Dominica, Argentina, Reunion, Barbados, Hawaii, Mexico, Puerto Rico, Philippines, Bangladesh, Mauritius, and Peru, etc., all of which are directly developed by the above parents. F134 was developed by crossing POJ2878 with Co290 in Taiwan Province of China, and CGT11 and CYZ 81-173 were developed by crossing Co419 with CP49-50 in the Sugarcane Research Institute of Guangxi Academy of

Agricultural Sciences and Yunnan Academy of Agricultural Sciences. At present, almost all commercial varieties in the world have the kinship of these varieties.

Due to the poor resistance and narrow adaptability of *S. officinarum*, the value of *S. spontaneum* inbreeding is mainly manifested in the adaptability to drought, low temperature, barrenness, salinity, swamp, etc. Through the continuous input of *S. spontaneum*, the cultivation range of sugarcane extends to 38°N (such as Spain) and 33°S (Australia), and China also reaches 33°N (Hanzhong, Shanxi), close to the edge of the northern boundary; The vertical distribution of sugarcane is constantly challenging new heights. The altitude of sugarcane planting ranges from > 1,300 m to 1,400 m and 1,600 m, and some sugarcane areas even reach 1,800 m. The results of sugarcane sexual hybridization breeding on the use of *S. spontaneum* make sugarcane breeders confident in improving sugarcane varieties by using wild germplasm. The use of *S. spontaneum* has become an important part of sugarcane variety improvement and development of breakthrough variety in the world, including Brazil, India, China, Australia, Barbados, India, the United States, and Thailand (Huang, 1983; Wang et al., 2006; Wu et al., 2011). There are two main ways to utilize *S. spontaneum* in sugarcane cross breeding: *S. officinarum* × *S. spontaneum* and *Saccharum* complex × *S. spontaneum*. Different ways of utilization have different breeding achievements.

3.2.1.1 Hybridization mode and effect of S.*officinarum* × *S. spontaneum*

In 1890, sugarcane breeder Jeswei discovered a remarkable clone of *S. spontaneum* in a mountain in Java, taller, juicier than the local wild species (*S. spontaneum*). Although the sugar content is lower than that of the *S. officinarum*, it is much higher than that of *S. spontaneum*. Through careful analysis, it is considered that this clone has the characteristics of both *S. officinarum* and *S. spontaneum*. The natural hybrid, Black cheribon/ *S. officinarum* × Glagah/ *S. spontaneum*, is named *Kassoer*, which means hybrid superiority. The post-cytologist Bremer, G., 1924, confirmed that Jeswei's analysis was correct. This discovery enhanced the confidence of Jeswei and later sugarcane breeders in improving sugarcane by using *S. spontaneum*. Thus, the breeding theory of "Nobilization" of sugarcane was formed; that is, *S. officinarum* was used as a female parent to cross closely with *S. spontaneum*, and continuous backcross utilization was used to sugarcane varieties more and more excellent. The breeding of POJ2878 is a symbolic event of "noble" breeding and a significant breakthrough in sugarcane breeding in the world. Later, sugarcane breeding in the world crossed under *S. officinarum* × *S. spontaneum* and successfully obtained noble offspring such as Co205, Co206 (Vellai × S.Spont/India *S. spontaneum*), and Co291 (K.Boothan × S.spont/India *S. spontaneum*). The first noble variety of *S. officinarum* × *S. spontaneum* cross only shows advantages in sugarcane stalk and sucrose yield but also shows strong stress resistance, tolerance to extensive cultivation, drought tolerance, barren resistance, and cold tolerance, developed root system, tillering solid ability, good ratooning, early maturity, high sugar content and high fiber content, which is called "wonderful cane" and is also an interspecific hybridization of sugarcane. After cytological and chromosomal analysis, it is considered that *S. barberi* ($2n = 82-124$) and *S. sinense* ($2n = 116-118$) are also the natural hybrid offspring of *S. officinarum* and *S. spontaneum*. In the world of sugarcane cross-breeding, the *S. officinarum* × *S. spontaneum* cross was successfully carried out. The institutions that bred excellent varieties from their offspring were Java in Indonesia, Coimbatore in India, Mauritius, and Sugarcane Breeding Station in Hainan (HSBS) in China (Table 3-9).

Table 3-9 Hybridization of *S. officinarum* × *S. spontaneum*

Variety/Clone	Parentage	Location	Cultivar developed
Kassoer	Black cheribon × Glagah	Java, Indonesia	Many, including POJ2878 and so on
Co205, Co206	Vellai × India *S. spontaneum*	Coimbatore, India	Many, including CP49-50 and NCo310
Uba Marot	*S. officinarum* × *S. spontaneum*	Unknown	Many, V114, R541

3 The Hybrid Mode and Effect of Sugarcane Wild Resources

continued

Variety/Clone	Parentage	Location	Cultivar developed
Co291	K.Boothan × India *S. spontaneum*	Coimbatore, India	Many, including Co290, etc
Co285	S.Mauritius × India *S. spontaneum*	Coimbatore, India	Many, including NCo310
M4500, M4600	Saretah × *S. spontaneum*	Mauritius	Many, Co331, CB68-41
M2	K.Boothan × India *S. spontaneum*	Mauritius	No reports
M1515	Nanaal × India *S. spontaneum*	Mauritius	No reports
CYC 58-47	Badila × Yacheng *S. spontaneum*	Yacheng, Hainan	Many, CYC 71-374 and so on
CYC 58-43	Badila × Yacheng *S. spontaneum*	Yacheng, Hainan	Many, CGTNo.17 and so on
CYC 82-108	Badila × Yunnan *S. spontaneum*	Yacheng, Hainan	Many, CGT29 and so on
CYC 75-419	Fiji × Yacheng *S. spontaneum*	Yacheng, Hainan	No
CYC 79-290	Crystalina × Lingshui *S. spontaneum*	Yacheng, Hainan	No
CYC 95-29	Badila × Yacheng *S. spontaneum*	Yacheng, Hainan	No

Source: Wu et al., 2014.

Since the founding of the Hainan Sugarcane Breeding Station, China (HSBS) has researched broadening sugarcane's genetic basis using wild germplasm (Huang, 1983; 1991) and hybrid utilization of *S. spontaneum* is the most effective. The collected *S. officinarum* such as Badila, Fiji, and Black Cheribon is used as the starting materials for "noble" breeding. A batch of excellent breeding materials have been screened out, and a large number of "Yacheng" series of excellent sugarcane breeding new materials with the genetic background of *S. spontaneum*, such as CYC 58-47, CYC 58-43, and CYC 82-108, have been bred by various breeding institutions in China. Including CYT 64-395, CYT 85-881, CYT 89-113, CZZ 82-339, CYN75-191, CYN89-759, CGT13, CGT14, CGT17, CGT21, CGT96-44, and CCT89-103 (Deng et al., 1996; Deng et al., 2007). CGZ 18 (CGN 95-108) bred by Jiangxi Sugarcane Research Institute and CGT21 bred by Guangxi Sugarcane

Research Institute (GXSRI) in recent years are all descendants of Badila (Deng et al., 2007). Sugarcane Research Institute of Yunnan Academy of Agricultural Sciences (YSRI) has successfully selected a batch of combinations (Table 3-10) by crossing different *S. officinarum* from different sources and obtained a large number of progenies with excellent performance such as good stool shape, many tillers and strong ratooning (Wang et al., 2006). Due to late cross-utilization, the offspring have excellent performance. Still, it is necessary to continue to increase the intensity of reciprocal cross-utilization to make it work for the development of the sugarcane industry as soon as possible.

Table 3-10 New attempt of hybridization of *S. officinarum* × *S. spontaneum*

Cross combination	Type	Hybridization type	Location
Muckche × Guandong 4	F_1	*S. officinarum* × *S. spontaneum*	Kaiyuan, Yunnan
Barwilspt × Hainan 92-63	F_1	*S. officinarum* × *S. spontaneum*	Kaiyuan, Yunnan
Luohancane × Hainan 92-9	F_1	*S. officinarum* × *S. spontaneum*	Kaiyuan, Yunnan
Luohancane × Hainan 92-8	F_1	*S. officinarum* × *S. spontaneum*	Kaiyuan, Yunnan
Luohancane × Guandong10	F_1	*S. officinarum* × *S. spontaneum*	Kaiyuan, Yunnan
Luohancane × Vietnam.3	F_1	*S. officinarum* × *S. spontaneum*	Kaiyuan, Yunnan

Source: Wu et al., 2014.

3.2.1.2 Hybridization mode and effect of *Saccharum* complex × *S. spontaneum*

Saccharum complex × *S. spontaneum* is one of the primary cross-utilization ways of *S. spontaneum*. Table 3-11 lists 27 basic cross combinations of *Saccharum* complex and *S. spontaneum* and many different types of offspring which have been screened out. Usually, hybrids have excellent characters but poor resistance after many crosses. Crossing with *S. spontaneum* usually improves hybrids by using wild species' resistance to gain more resistance and adaptability. Mauritius and Barbados used B6308 and Ba11569 to hybridize with Indian *S. spontaneum* and developed Mo1928 and B37256, and Hawaii used Mo1928 to hybridize with UD50 and developed H33-6705; HSBS crossed K28, POJ2878, and CYC 60-75 to different types of *S. spontaneum*. Although *S. spontaneum* CYC 55-7, CYC 75-411, CYC 82-110, and CYC 85-45 of the F_1 generation were successfully obtained, there were no reports of cultivation of sugarcane varieties. RSBS and YSRI also carried out a large number of hybridization with *S. spontaneum* (Wang et al., 2006; Chu, 2000; Tao, 1996; Yang, 2004), using the *S. spontaneum* in Yunnan, Xishuangbanna, Gaoligong Mountain, Binchuan, etc., divided into humid tropic and dry tropic types, and divided into drought-tolerant, cold-tolerant, disease-resistant and barren-tolerant types in terms of tolerance, and developed a large number of clones with different ecological types of *S. spontaneum* in Yunnan (Table 3-11). Their offspring have reached by providing productive hybrid tassels to the whole country. It is generally reflected that the offspring have vital permanence, many stalks, and high yield, but whether it can cultivate excellent varieties still needs practical test. Analysis of Table 3-11 showed that among the 27 cross combinations, 16 genotypes of *Saccharum* complex and 20 of *S. spontaneu*m are used. Among the 16 Saccharum complex, there are many world-class excellent parents or germplasms, such as EK28, POJ2878, Co419, etc. Still, the offspring's performance is not as expected; most of them have not developed cultivar, only three cross combinations developed cultivar, and their adaptability is poor. Their popularization area is limited, so they did not become breakthrough varieties, which breeders deserve deep consideration.

Table 3-11 Status of *Saccharum* complex × *S. spontaneum*

Number	Variety / Clone	Cross combination	Note	Progeny
1	B37256	Ba11569 × Indian *S. spontaneum*	Barbados, China	More
2	Mo1928	B6308 × Indian *S.spontaneum*	Mauritius, China	H33-6705
3	CYR 80-189	POJ3016 × Manghao *S. spontaneum*	Ruili, Yunnan, China	CYZ 99-155
4	1047S3	POJ2940 × Java *S. spontaneum*	Japan. Saiban, China	No
5	CYC 55-7	POJ2878 × Yacheng *S. spontaneum*	Yacheng, Hainan, China	No

				continued
Number	Variety / Clone	Cross combination	Note	Progeny
6	CYC 75-411	EK28 × Yacheng S. spontaneum	Yacheng, Hainan, China	No
7	CYC 82-110	EK28 × Yuanjiang S. spontaneum	Yacheng, Hainan, China	No
8	CYC 85-45	CYC 60-75 × Yunnan S. spontaneum	Yacheng, Hainan, China	No
9	CYR 80-15	F134 × Manghao S. spontaneum	Ruili, Yunnan, China	No
10	CYR 80-161	POJ3016 × Yunnan S. spontaneum	Ruili, Yunnan, China	No
11	CYR 91-3406	CYR 80-189 × Yunnan S. spontaneum 83-255	Ruili, Yunnan, China	No
12	CYR 92-81	CCT 61-480 × Yunnan S. spontaneum 83-157	Ruili, Yunnan, China	No
13	CYR 92-114, CYR 92-128, CYR 92-148, CYR 93-28	CGZ 14 × Yunnan S. spontaneum 83-157+82-114	Ruili, Yunnan, China	No
14	CYR 93-36	CGZ 14 × Yunnan S. spontaneum 83-157	Ruili, Yunnan, China	No
15	CYR 96-119 CYR 96-124	CYR 93-2418 × Yunnan S. spontaneum 114	Ruili, Yunnan, China	No
16	CYR 97-105	CYR 91-2868 × Yunnan S. spontaneum 114	Ruili, Yunnan, China	No
17	CYR 96-124	CYR 93-2418 × Yunnan S. spontaneum 80-114	Ruili, Yunnan, China	No
18	CYR 00-298	CGZ 14 ×Yunnan S. spontaneum 82-48	Ruili, Yunnan, China	No
19	CYZ 99-3	CGZ 14 × Vietnam S. spontaneum 2	Kaiyuan, Yunnan, China	No
20	CYZ 02-356	Co419 ×Yunnan S. spontaneum 75-1-2	Kaiyuan, Yunnan, China	No
21	CYZ 00-57, CYZ 00-63, CYZ 00-77	Co617 ×Yunnan S. spontaneum 1	Kaiyuan, Yunnan, China	No

continued

Number	Variety / Clone	Cross combination	Note	Progeny
22	CYZ 00-271, CYZ 00-287	CGZ 14 × Vietnam S. spontaneum 3	Kaiyuan, Yunnan, China	No
23	CYZ 00-339, CYZ 00-347, CYZ 00-350	CGZ14 × Hainan S. spontaneum 92-66	Kaiyuan, Yunnan, China	No
24	CYZ 00-374	CGZ 14 × Hainan S. spontaneum 92-9	Kaiyuan, Yunnan, China	No
25	CYZ 00-320	CGZ14 × Yunnan S. spontaneum 87-49	Kaiyuan, Yunnan, China	No
26	CYZ 00-202	Co419 × Guangdong S. spontaneum 81	Kaiyuan, Yunnan, China	No

Source: Wu et al., 2014.

The research shows that:

Firstly, there are many varieties developed by the hybridization model *S. officinarum* × *S. spontaneum*; these varieties were significant breakthroughs, wide adaptability, and sizeable growing area, and have made an outstanding contribution to the sugar industry in the world. Further, it is found that all varieties with the significant breakthrough in offspring are developed through hybridization within the same or equal generations, and the most prosperous countries are Indonesia, India, and China.

Secondly, there are significant differences in the characteristics of developed cultivars due to the different utilization modes of generation or backcross. The hybridization conducted within the same or equal generations may generate a breakthrough in offspring populations. For example, F134, Co419, NCo310, CP49-50, and CGT11 are breakthrough varieties produced by hybridization within the same or equal generations. If unequal backcross is used, even the parental clone themselves are outstanding; it may difficult to develop elite new sugarcane cultivars. For example, the hybridization between Co290 and CP49-50 is also asymmetric, and the varieties Gannan 61-451(CGN61-451) and CGT60-149 (CGT3) cultivated in

China but had not been widely grown.

Thirdly, in the *S. officinarum*× *S. spontaneum* hybridization mode, the difference in selected parental clones impact the performance of progeny. Some progeny become worldwide varieties or parents, while others fail in developing any variety. Therefore, it is vital to select *S. officinarum* and *S. spontaneum* with excellent characters as basic parental clones in the hybridization.

Fourthly, in recent years, there are a large number of cross combinations between *S. officinarum* and *S. spontaneum*, but the number of newly used *S. officinarum* is small, only Badila, Muckche, Barwilspt, and Luohancane. Because of the short hybridization time, there is no report on developing new cultivars from the latter three *S. officinarum*. If all of them are used for peer-to-peer hybridization, only two pairs of F_1 progeny with complementary characters and no genetic cross can be found, and then peer-to-peer hybridization is only F_2 generation, and at least F_3 generation is needed for breeding breakthrough varieties.

Fifthly, *Saccharum* complex× *S. spontaneum* has early utilization time, many types and progeny, but there are almost no excellent derivative varieties in its offspring and derived offspring. Although some varieties have been released, their adaptability is poor, their performance is unstable, and their popularization area is limited, which should be avoided as far as possible during hybridization.

3.2.2 The Way and Effect of Hybrid Utilization of *S. robustum*

The major contribution of *S. robustum* to sugarcane breeding lies in the improvement of lodging resistance, drought tolerance, and ratooning ability. As a wild resource in sugarcane, the main breeding value of *S. robustum* is that the offspring show great vigor, the sugarcane stalk is as hard as bamboo, with solid wind tolerance, insect resistance, drought tolerance, developed root system, good ratooning, and high fiber content (Luo, 1984; Wang, 1979), as a crucial wild resource of sugarcane cross-breeding, has made essential contributions to sugarcane breeding. With the development of science and technology and social production, sugarcane production accelerates the transition from artificial to mechanized. The lodging resistance of sugarcane varieties will be paid more and more attention to sugarcane breeding. At present, the reported utilization modes of *S. robustum* mainly include *S. officinarum* × *S. robustum*, *S. robustum* × *S. officinarum,* and *Saccharum* complex × *S. robustum*.

3.2.2.1 The hybridization mode and effect of *S. officinarum* × *S. robustum*

S. officinarum × *S. robustum* has been selected for many cross combinations (Table 3-12), but there are no reports indicated the success in developing a new cultivar. Since the mid-1970s, the cross combinations Badila×*S. robustum*, 50Uahiapele×*S. robustum* and Badila× *S. robustum*/57NJ208 have been used in developing CYC 75-280, CYC 95-4, CYC 96-48, CYC 96-37 and CYC 97-39, etc. RSBS successfully crossed the *S. officinarum* 48Mouna with *S. robustum* 57NG208 and obtained F_1 generation CYR 03-229. From 1998 to 2002, YSRI developed BC_1 of some *S. robustum* and *S. officinarum* by backcrossing CYC 96-48 with commercial variety. Among them, excellent germplasm materials such as CYZ 2000-505, CYZ 2000-506, and CYZ 2000-530 have great vigor, medium to large stalk and number of millable stalk reaching 125,100−150,100 per hectare, high cane yield, and high sugar content (14.09%−14.67%). CYZ 2002-105, CYZ 2002-107, CYZ 2002-112, CYZ 2002-115, and other clones have a brix of 23.0%−24.4% in January (Wang et al., 2003).

3.2.2.2 Hybridization mode and effect of *S. robustum* × *S. officinarum*

Since the 1970s, Hainan Sugarcane Breeding Station, Institute of Sugarcane and Sugar Science, Ministry of Light Industry (HSBS), and selected reciprocal crosses *S. robustum* × *S. officinarum* (Table 3-12). Progenies such as CYC 75-270 and CYC 75-273 were advanced. Although no variety was developed, a beneficial exploration has been made in the utilization of *S. robustum*.

Table 3-12 Cross utilization between *S. officinarum* and *S. robustum*

Variety/Clone	Parentage	Hybridization mode	Location
CYC 75-280, CYC 75-281	Badila × *S. robustum*	*S. officinarum* ×*S. robustum*	Yacheng, Hainan
CYC 95-4, CYC 96-48	Badila × 57NJ208	*S. officinarum* ×*S. robustum*	Yacheng, Hainan
CYC 96-37, CYC 97-39	50Uahiapele × *S. robustum*	*S. officinarum* ×*S. robustum*	Yacheng, Hainan
CYR 03-229	48Mouna × 57NG208	*S. officinarum* ×*S. robustum*	Ruili, Yunnan
CYC 75-270, CYC 75-273	*S. robustum* × Badila	*S. robustum* ×*S. officinarum*	Yacheng, Hainan
CYC 97-46	NG77-1 × Ganapathy	*S. robustum* ×*S. officinarum*	Yacheng, Hainan

Source: Wu et al., 2014.

3.2.2.3 Hybridization mode and effect of *Sugarcane* complex × *S. robustum*

There is a long history of crossing the *Sugarcane* complex with *S. robustum* (Table 3-13). In 1932, Hawaii began to cross Mol1231 to *S. robustum*s, and the F_1 generation H32-5774, then F_2 generation H34-1874 (H27-8101×H32-5774), F_3 generation H37-1933 (H32-8560×H34-1874); Queensland developed 32MQ, USDA ARS Sugarcane Field Station at Canal Point developed CP36 and other varieties. The sugarcane breeding in Taiwan Province of China has achieved great success in the cross-utilization of the *Sugarcane* complex × *S. robustum*. In 1939, the combination POJ2883×*S. robustum* was used to develop PT39-461; Use POJ2725 as a female parent, develop PT40-203, and use POJ2875 as a female parent to developed PT40-345. PT40-388 was developed with F108 as a female parent (Luo, 1984). The most effective hybrid combination is POJ2875 × *S. robustum*, from which PT40-345 was developed. after backcrossing with POJ2878 (POJ2878 × PT40-345) PT43-52 was developed. In 1946, it was found that this clone has strong wind resistance, and a batch of excellent lodging resistance has been developed by using PT43-52, Many F and ROC, such as F146, F152, F160, F172, ROC1, ROC5, ROC10, ROC16, ROC22, etc., are offspring of PT43-52. These varieties have been widely grown in

China, and they are also essential parents commonly used by breeding institutions in China. ROC16 and ROC22 have been widely cultivated in China.

Hainan Sugarcane Breeding Station, Institute of Sugarcane Sugar Science, Ministry of Light Industry (HSBS), obtained fuzzy from combination S17 × *S. robustum* 1980s, and developed CYC 80-142 CYC 80-143. CYC 96-49 (IJ76-315 × NG77-004/*S. robustum*) was obtained by crossing IJ76-315 with *S. robustum* NG77-004 in the 1990s. However, no report indicating they succeeded in developing sugarcane cultivar. Besides, Guangzhou Sugarcane Industry Research Institute (GZSRI) has developed excellent parent clones such as CYC 73-226 and CZZ 80-101 using the progeny of *S. robustum* PT40-388 and developed large-scale extension varieties such as CYT 79-177 (Deng et al., 2004). In recent years, the cultivar Mingtang 70-611 and *S. robustum* 57NG208 have been successfully crossed in Ruili Sugarcane Breeding Station in China (RSBS), F_1 generation CYR03-403 (CMT 70-611 × *S. robustum*/57NG208) was obtained. After breeding excellent germplasm/parents, it is expected to breed new sugarcane varieties with strong lodging resistance and mechanized cultivation adaptability.

Table 3-13 Hybridization of sugarcane hybrid × *S. robustum*

Variety/Clone	Cross (sugarcane hybrid × *S. robustum*)	Location	Progeny
H32-5774	Mol1231 × Mol123 (*S. robustum*)	Hawaii, USA	No
PT39-461	POJ2883 × *S. robustum*	Taiwan, China	No
PT40-203	POJ2725 × *S. robustum*	Taiwan, China	No
PT40-345	POJ2875 × *S. robustum*	Taiwan, China	Many (PT43-52 etc.)
PT40-388	F108 × *S. robustum*	Taiwan, China	Many (CYT 79-177, etc.)
CYC 96-49	IJ76-315 × NG77-004 (*S. robustum*)	Yacheng, Hainan, China	No
CYC 80-142, CYC 80-143	S17 × *S. robustum*	Yacheng, Hainan, China	No

			continued
Variety/ Clone	Cross (sugarcane hybrid × S. robustum)	Location	Progeny
CYR 03-403	CMT 70-611 × 57NG208 (S. robustum)	Ruili, Yunnan, China	No

Source: Wu et al., 2014.

The results show that: There have been a large number of hybridization conducted under the models of S. officinarum × S. robustum and S. robustum × S. officinarum since the 1970s, and have also developed elite clones, but no contribution to the sugarcane genetic improvement has been made through hybridization with S. robustum so far, so it is necessary to continue further improvement on the hybrids obtained after 1990s; Sugarcane complex × S. robustums make a more significant contribution than original S. officinarum × S. robustum. It has been used early (since the 1930s), and many types have been involved, and its descendants have made outstanding contributions to sugarcane breeding, especially the combination of POJ2875× S. robustum. After PT43-52 was developed, many varieties with good wind resistance and strong adaptability were developed from its derived offspring, and many F and ROC varieties were its descendants.

References

BESSE P, MECINTYPE C L, BERDING, 1996. Ribosomal DNA variations in *Erianthus*, a wild sugarcane relative (Andropgoneae-Saccharinae)[J]. Theoretical and Apply Genetics, 92 (6):733-743.

BRANDES E W, SARTORIS G B, 1936. Sugarcane: Its origin and improvement[R]. USDA. Year Book.

CHEN R K, LIN Y Q, ZHANG M Q, et al., 2003. Theory and practice of modern sugarcane breeding[M]. Beijing: China agriculture press (in Chinese).

CAI Q, FAN Y H, 2005. Systematic evolution and genetic relationship of sugarcane complex by AFLP[J]. Acta Crops Sinica, 31(5):551-559.

CHEN X W, DENG H H, CHEN Y S, 2010. Utilization of Badila in the breeding of YC-Series Parents and new varieties of sugarcane[J]. Sugarcane and canesugar (6):1-5.

DENG H H, LI Q W, CHEN Z Y, 2004. Innovation and utilization of sugarcane parents[J]. Sugarcane,11 (3): 7-12.

DENG Z H, LI Y C, LIU W R, et al., 2007. Chromosome genetic analysis for the hybrid progeny of *S. officinarum* L. and Erianthus arundinaceum[J]. Chinese Journal of Tropical Crops, 28 (3):62-67.

D'HONT A, PAULET F, GLASZMANN J C, 2002. Oligoclonal interspecific origin of 'North Indian' and 'Chinese' sugarcanes[J]. Chromosome Research (10):253-262.

FU C, DENG H H, CHENG X W, 2003. Research and utilization of *Erianthus arundinaceus* at Hainan sugarcane breeding station[J]. Sugarcane and Canesugarr (6):1-14.

FU C, LIU S M, YANG Y H, et al., 2004. Review on research and utilization of sugarcane germplasm in Hainan sugarcane breeding station during the Ninth Five-Year Plan[J]. Sugarcane and Canesugar (3):1-8.

GRASSL C O, 1967. Introgression between *Saccharum* and *Miscathus* in New Guinea and the Pacific area[J]. Proc. ISSCT (12):995-1003.

JING Y F, TAO L A, YANG L H, et al. 2006. Preliminary study on isozyme identification of parent-child relationship of sugarcane hybrid offspring[J]. Sugarcane and Canesugar (6):1-4.

HUANG J Y, LIAO J X, ZHU G Y, 1997. Intergeneric copulatality of *Saccharum* L. with *Narenga porphyrocoma*, *Miscanthus Floridulus* and *Erianthus kockii*, the morphology and isozyme analysis of their hybrid F_1 clones[J]. Southwest China Journal of Agricultural Sciences, 10 (3):92-96.

HUANG Q Y, 1991. Viewing the breeding ways of China and the contribution of Hainan Sugarcane Breeding Station from the new sugarcane varieties bred in 1980s[J]. Sugarcane and Canesugarr (4):7-14.

LI F S, HE L L, YANG Q H, et al., 2003. Evaluation of some special characters of

Erianthus fulvus and Identification of sugarcane hybrids based on chromosome number and RAPD[J]. Molecular Plant Breeding (5/6):775-781.

LI X W, LUO J S, 1948. Cell Studies on Sugarcane Plants-Noble Species, Thatched Grass and Wild Hybrid[J]. Botanical Report of National Academia Sinica, 1948 (2): 147-160.

LI Y R, 2010. Modern sugarcane science[M]. Beijing: China Agricultural Press (in Chinese).

LIU H B, 1997. Research and utilization of sugarcane distant hybridization in mainland China[J]. Sugarcane, 4 (3):7-9.

LIU S M, 1992. Utilization Effect of Several *Erianthus arundinaceus* Offspring as Sugarcane Hybrid Parents[J]. Sugarcane and Canesugarr (3):1-6.

LIU S M, FU C, CHEN Y S, 2007. Utilization of *Erianthus arundinaceus* progeny through backcross over the last decade at Hainan sugarcane breeding station[J]. Sugarcane and Canesugarr (4):1-6.

LIU S M, WANG L N, HUANG Z X, et al., 2011. Utilization of sugarcane parents of Yacheng series in sugarcane breeding of China[J]. Sugarcane and Canesugarr, (4):5-10.

LUO J S, 1984. sugarcane science[C]. Guangzhou: Guangdong sugarcane society.

PENG S G, 1990. Sugarcane Breeding[M]. Beijing: Agricultural Press (in Chinese).

PIPERIDIS G, D' HONT A, 2001. Chromosome composition analysis of various Saccharum interspecific hybrids by genomic in situ hybridization (GISH) [J]. International Society Sugar Cane Technologists (24):565-566.

SHEN W K, 2002. Discussion on the value of hybrid utilization of *Erianthus arundinaceus*[J]. Sugarcane, 9 (3):1-5.

SOBHAKUMARI V P, 2005. Chromosome numbers of some noble sugarcane clones[J]. Journal of Cytology and Genetics (6):25-29.

STEVENSON G C, 1965. Genetics and breeding of sugarcane[M]. London: Longmans, Green and Co.

TAO L A, 1996. Study on heterosis and utilization of wild sugarcane resources in Yunnan[J]. Yunnan Agricultural Science and Technology (5):13-16.

TAO L A, JING Y F DONG L H, et al., 2011. Utilization and screening of interspecific hybrid combinations of sugarcane cultivars.

TEW T L, WU K K, SCHNELL R J, et al., 2011. Comparison of biparental and melting pot methods of crossing sugarcane in Hawaii[J]. Sugar Tech,12(2):139-144.

WANG J M, 1985. Sugarcane Cultivation in China[M]. Beijing: Agricultural Press (in Chinese).

WANG L P, CAI Q, LU X, et al., 2008. Study of wild species *Erianthus rockii* germplasm innovation and use[J]. Sugar Crops of China (2):8-11.

WANG L P, CAI Q, FAN Y H, et al., 2007. Study on distant hybridization between sugarcane and *Erianthus arundinaceus*[J]. Southwest China Journal of Agricultural Sciences, 20 (4):721-726.

WANG L P, MA L, XIA H M, et al., 2006. Utilization of *S. spontaneum* in cross breeding[J]. Sugar Crops of China (1):1-4.

WANG L P, FAN Y H, CAI Q, et al., 2003. Research progress on hybrid utilization of sugarcane germplasm resources[J]. Sugarcane (3):7-23.

WANG Z L, WANG S Q, PAN S M, et al., 1999. Study on superior sugarcane germplasm resources and their utilization in breeding[J]. Sugar Crops of China (4):12-15.

WEN Y, 1998. Intergeneric hybridization and chromosome behavior of sugarcane related plants[J]. Sugarcane and Canesugarr (3):1-7, 17.

WU N Y, QI J W, 1987. Identification of distant hybrids between sugarcane and *Leymus chinensis*[J]. Journal of South China Agricultural University, 8 (2):28-34.

WU C W, 2005. Discussion on germplasm innovation and breeding breakthrough varieties in sugarcane[J]. Southwest China Journal of Agricultural Sciences, 17 (6):858-861.

WU C W, PHILLIP J, LIU J Y, et al., 2011. Inheritance of quality traits of the distant crossing between *S. officinarum* and *S. spontaneum*[J]. Journal of plant genetic resources, 12 (1): 59-63.

WU C W, ZHAO P F, XIA H M, et al., 2014. Modern cross breeding and selection techniques in sugarcane[M]. BeiJing: Science Press (in Chinese).

WU Z K, CHEN Y, 2011. Proceedings of the 14th Symposium of Sugarcane Professional Committee of China Crop Society in 2011 (Yunnan Volume)[C]. Kaiyuan City, Yunnan Sugarcane Association: 18-23.

YANG L H, 2004. Studies on posterity of *Saccharum spontaneum L.* and *S.Arundinaceum* and *Erianthus rocki*i for resistance to smut in Yunnan[J]. Sugarcane, 11(1):10-14.

ZHANG M Q, DENG Z H, CHEN R K, et al., 2006. Genetic improvement and efficient breeding of sugar crops[M]. Beijing: China Agricultural Press (in Chinese).

ZHOU Y H, HUANG H N, YIN P J, et al., 1997. Study on sugarcane germplasm in Hainan Sugarcane Breeding Station (ii)[J]. Sugarcane and Canesugarr (3):2-7.

LIANG Z G, HUANG J S, 1996. Preliminary report on hybridization of sugarcane and its related wild plants[J]. Guangxi Agricultural Sciences (1):5-6.

4 The Hybrid Mode and Effect of Original Saccharum Cultivated Species

Saccharum L., including *S. officinarum, S. barberi,* and *S. sinense*, is the essential commercial trait gene resource in sugarcane breeding. It has been used in production as a cultivated variety in history. Since the hybridization of sugarcane, the original parent has the most hybridization times and the most significant utilization. It has also made the most outstanding contribution to sugarcane breeding, in the process of "noble" breeding of sugarcane, the original of important economic characters such as high yield, high sugar, big stalk, juicy, etc. At present, the excellent varieties for promotion in production are the hybrids of one, two, or three of the above-mentioned original species. Most modern sugarcane varieties contain the genetic background of *S. officinarum, S. barberi,* and *S. sinense*, accounting for about 90%, and other related genera accounting for about 10% of the genetic background.

4.1 The Hybrid Utilization Mode and Effect of *S.officinarum*

S. officinarum, also known as noble cane, is the most important genetic resource of modern sugarcane cultivar. There are five hybridization modes of *S. officinarum*, including *S. officinarum* × *S. officinarum, S. officinarum* × *S. sinense, S. officinarum* × *S. barberi, S. officinarum* × *S. spontaneum, S. officinarum* × *S. robustum*, which are the most widely used among various utilization modes at present. The earliest sugarcane breeding is conducted by selection from seedlings generated from naturally pollinated *S. officinarum*. For example, in Java, the EK series was developed by Otaheiti. Lahaina×Fiji developed the EK2. Crystalina and White Transparent developed the D74; the H109 (Lahaina × US1494) was developed in Hawaii, becoming a breakthrough in sugarcane scientific research. Because *S. officinarum* and their offspring only adapt to the specific tropical environment and are vulnerable to diseases and insects, breeders began to consider broadening the genetic basis of *S. officinarum* to improve their adaptability and enhance their disease resistance and insect resistance (Stevenson, 1965). Therefore, the sugarcane "nobilization" breeding was produced; that is, the *S. officinarum* was used as the female parent to cross closely with the *S. spontaneum*, the continuous backcross with the *S. officinarum*. Since the cross-breeding of sugarcane, the theory of "nobilization" has been expanding, and the cross between *S. officinarum* and *S. sinense*, *S. barberi*, *S. robustum*, and related wild species has been coming out. The cultivation scope of sugarcane has been expanding, and breakthroughs have been made in planting latitude and altitude, promoting the sugar industry's continuous development.

Table 4-1 Basic hybridization of *S. officinarum* × *S. officinarum*

Variety/Clone	Cross combination	Location	Progeny
D74	Crystalina × W.Transparent	Demerara, Guyana	many
POJ100	Loethers × B.Bangjermasin hitan	Java, Indonesia	many
EK2	Lahaina × Fiji	Java, Indonesia	many
SW3, SW11, DI46, DI52	B.Cheribon × Batjan	Java, Indonesia	many
H27-8101	Badila × H25C14	Hawaii, USA	many
H109	Lahaina × US1494	Hawaii, USA	many
B247	B.Cheribon × Fiji	Barbados	No
CYC 76-9	Lethers × B.cheribon	Yacheng, Hainan, China	No

Source: Wu et al., 2014.

4.1.1 Hybridization between *S. officinarum*

The earliest cross between *S. officinarum* was carried out under natural conditions, first discovered by Dutchman Sotwedel (Sotwedel, 1887). Later, Harrison et al. (1888) discovered them in Java at 10°S and Barbados at 10°N, respectively, so the first hybrids bred were natural hybrids among cultivated species. To use more and better *S. officinarum* for developing sugarcane varieties needed by human beings, Java sugarcane breeder Jesweit first started the artificial cross-breeding of sugarcane and put forward the breeding goal of noble sugarcane. Kobus (1893–1905) in Java first pollinated with cage method and first developed the "noble" generation POJ 100 (Loethers × Black Bangjermasin Hitan). At the beginning of the 20th century, the cross between *S. officinarum* once prevailed. The *S. officinarum*× *S. officinarum* hybridization is successfully realized. Excellent successful examples are D74 (Crystalina ×White Transparent), B247 (Black Cheribon × Fiji), EK2 (Lahaina × Fiji), H109 (Lahaina × US1494), H27-8101 (Badila × H25C14), POJ100 (Black Bangjermasin Hitan × Loethers), SW3, SW11 and DI46, DI52 (Black Cheribon × Batjan), etc. India used D74 (F_1) to continue backcrossing Co221(F_2) and bred BC_2 generation of *S. officinarum* Co290 (Co 221 × D74). In Java, EK28 was developed from the hybridization of two F_1 (EK2 and POJ100) of *S. officinarum*, containing the genetic background of 4 *S. officinarum*. The hybridization between POJ100 and Kassoer created an F_2 progeny POJ2364, and the hybridization between POJ2364 and EK28 (hybridization between equal generation) created the F_3 progeny POJ2878. POJ2878 was crossed to Co290 and developed Co419. POJ2878, Co290 and Co419 are juxtaposed as the King of Sugarcane. They have excellent commercial, resistance, and adaptability, are widely used as economic varieties in sugarcane areas and have good flowering characteristics and pollen fertility. They can be used as both male parents and female parents for cross-utilization. Many excellent varieties have been bred using them and their derived varieties for popularization and application (Luo, 1984; Peng, 1990).

The cross between *S. officinarum* has made an outstanding contribution to sugarcane breeding in the world. The progenies bred are not only excellent in character but also good parents. According to Peng (1990), there are about 711 *S. officinarum* genotypes, and only about ten have been used for cross breeding (Table 4-1), so there is still great potential for hybridization among *S. officinarum*. After the development of *S. officinarum* progenies such as D74, POJ100, EK2, SW3, and H27-8101, and their progeny varieties/parents were produced, Barbados successfully selected B.Cheribon×Fiji combination and bred innovative material B247; Hainan Sugarcane Breeding Station of Sugarcane Research Institute of the Ministry of Light Industry (HSBS) (Liu et al., 2011) also successfully selected and crossed the combination Lethers × B.cheribon, and screened out CYC 76-9; It is a pity that the *S. officinarum* used have been used in the original parents of POJ2878 and Co290, and then cross with the existing parent system, so they play a minor role in modern sugarcane cross-breeding. In recent years, Ruili Cross Breeding Base of Sugarcane Research Institute, Yunnan Academy of Agricultural Sciences (RSBS) began to cross new *S. officinarum*, and successfully selected 4 combinations of *S. officinarum*× *S. officinarum* (Table 4-2), and obtained more seedlings, all of which were true hybrids after authenticity identification (Tao et al., 2011), which is expected to make outstanding contributions to sugarcane breeding in the next step. The main problem is that only four new *S. officinarum* (Barwilspt, Pansahi, Zopilata, and Waie) were used in the four combinations.

Table 4-2 Utilization of new *S. officinarum*

Female parent	Male parent	Number of planted seedlings	Note
Barwilspt	Badila	350	Ruili, Yunnan
Barwilspt	Zopilata	425	Ruili, Yunnan
Barwilspt	Waie	19	Ruili, Yunnan
Pansahi	Zopitala	215	Ruili, Yunnan

Source: Tao et al., 2011.

4.1.2 Interspecific hybridization between *S. officinarum* and *S. sinense*, or landrace

S. sinense and Chinese landrace have been cultivated in China for a long time, are suitable for planting in a wide range, and have strong stress tolerance. They are the earliest cultivated varieties for sugar production in China. There are few reports of successful hybridization among *S. officinarum*, *S. sinense*, and landrace due to the difficulty of flowering, sterile pollen, and poor seedling performance. Due to the improvement of breeding conditions and hybridization technology, more successful hybridization reports among *S. officinarum*, *S. sinense,* and landrace. Hybridization among *S. officinarum*, *S. sinense,* and landrace is also a meaningful way to cultivate "noble" sugarcane varieties. According to current reports, there are four ways to cross *S. officinarum* with *S. sinense* and landrace (Table 4-3). In recent years, HSBS has successively carried out the cross between *S. officinarum*, *S. sinense,* and landrace, and developed CYC 97-20; RSBS produced fuzzy from 11 cross combinations of *S. officinarum*, *S. sinense* and landrace at one time, 3 combinations did not emerge, 1 combination was self-bred (900 seedlings), and 7 combinations produced 453 seedlings (Tao et al., 2011). The successful cross added the new genetic background to the current sugarcane breeding program.

Table 4-3 Hybridization among ***S.officinarum, S.sinense,*** and landrace

Cross combination	Hybridization type	Progeny	Location
Bailou cane × 50uahiapele	*S. sinense* × *S. officinarum*	CYC 97-24, CYC 97-25	Yacheng, Hainan
Ganapathy × Bailou cane	*S. officinarum* × *S. sinense*	CYC 97-47	Yacheng, Hainan
57NG155× Bailou cane	*S. officinarum* × *S. sinense*	CYC 98-27	Yacheng, Hainan
Nanjian chewing cane × Barwilsp	Local species × *S. officinarum*	Field planting of 0 seedlings	Ruili, Yunnan
Nanjian chewing cane × Marjar	Local species × *S. officinarum*	Field planting of 4 seedlings	Ruili, Yunnan
Canalana × Nanjian chewing cane	*S. officinarum* × Landrace	Field planting of 0 seedlings	Ruili, Yunnan
Zopitala × Nanjian chewing cane	*S. officinarum* × Landrace	Planting 900 seedlings (self-crossing)	Ruili, Yunnan
Badila × Henan Xuchang chewing cane	*S. officinarum* × *S. sinense*	Planted 5 seedlings	Ruili, Yunnan
Barwilspt × Henan Xuchang chewing cane	*S. officinarum* × *S. sinense*	Field planting of 400 seedlings	Ruili, Yunnan

continued

Cross combination	Hybridization type	Progeny	Location
Bawilspt × glossy bamboo cane	*S. officinarum* × *S. sinense*	Field planting of 29 seedlings	Ruili, Yunnan
Waie × Henan Xuchang chewing cane	*S. officinarum* × *S. sinense*	Field planting of 0 seedlings	Ruili, Yunnan
Guangze bamboo cane × Marjar	*S. sinense* × *S. officinarum*	Field planting of 1 seedling	Ruili, Yunnan
Henan Xuchang chewing cane × Badila	*S. sinense* × *S. officinarum*	Planted 3 seedlings	Ruili, Yunnan
Henan Xuchang chewing cane × Manjar	*S. sinense* × *S. officinarum*	Field planting of 11 seedlings	Ruili, Yunnan
Guangze bamboo cane × Henan Xuchang chewing cane	*S. sinense* × *S. sinense*	Field planting of 4 seedlings	Ruili, Yunnan

Note: Nanjian chewing cane is collected from Nanjian Yi Autonomous County, Yunnan Province, and its classification needs further confirmation.
Source: Wu et al., 2014.

4.1.3 Interspecific hybridization between *S. officinarum* and *S. barberi*

The hybridization between *S. officinarum* and *S. barberi* is early, with outstanding achievements, and the hybridization mode is *S. officinarum*× *S. barberi*. Chunnee is the only one *S. barberi* successfully used for hybridization, and other *S. barberi* are difficult to flower, so there is no report of their hybridization. In Java, the cross combination B.Cheribon× Chunnee was used for fuzzy production, and many excellent F_1 varieties such as POJ213, POJ143, and POJ181 were developed (Table 4-4). Among them, POJ213 has made the most outstanding contribution to sugarcane breeding in the world (Peng, 1990). Chunnee is the ancestor of the POJ, Co, B, H, and F series and some excellent varieties in China, such as NCo310, Co213, Co290, Co419, F134, and CGT11, etc. Meanwhile, the hybridization of G. Praenger × Chunnee generated cultivars POJ26, POJ33, POJ139, and POJ36, and the hybridization of ZW Cheribon × Chunnee generated a cultivar POJ1410. The offspring's contribution and their derived offspring of the latter two cross-crossed to sugarcane breeding were rarely reported.

Table 4-4 Basic hybridization utilization of *S. officinarum* × *S. barberi*

Cross combinations	Location	Cultivar developed	Progeny performance
B. Cheribon × Chunnee	Java, Indonesia	POJ213, POJ143, POJ181, POJ228, POJ234, POJ369, POJ826, POJ920, POJ979, POJ2379	There are many varieties developed
G. Praenger × Chunnee	Java, Indonesia	POJ26, POJ33, POJ36, POJ139	No further report
ZW Cheribon × Chunnee	Java, Indonesia	POJ1410	No further report

Source: Wu et al., 2014.

4.1.4 Cross between *S. officinarum* and sugarcane hybrids

The design of cross combinations generally requires that parents' traits complement each other, and the female parent's traits are better. The selected male parent may overcome the shortcomings of the female parent so that the offspring will be improved after hybridization, and the traits of commercial varieties are improved. *S. officinarum* has good commercial characters but poor adaptability and needs to grow well under much heat and water. However, sugarcane hybrids generally have the genetic background of *S. officinarum* and wild species in sugarcane, similar in genetic background. They have little difference in genetic basis, so hybridization is easy to conduct. There are two ways of crossing *S. officinarum* and hybrids: *S. officinarum* × hybrids or hybrids × *S. officinarum*. Cuba, India, Australia, Indonesia, Hainan, Ruili in China, and Pingtung in Taiwan Province of China all carry out many hybridizations (Table 4-5). Their offspring and derived offspring have successfully been used in developing new varieties.

Table 4-5 Hybridization between *S. officinarum* and hybrids

Variety/Clone	Cross combination	Type	Location	Progeny
25C14, 26C270	Y.Cheribon × H109	*S. officinarum* × hybrids	Cuba	H31-1389, H40-1179, H49-3646, US45-29-9, etc

continued

Variety/Clone	Cross combination	Type	Location	Progeny
Co214	S.Mauritius × M4600	*S. officinarum* × hybrids	Coimbatore, India	Co320, Co327, R331, CoTo, etc
POJ2221, POJ2222, POJ2320, POJ2191	B.Cheribon × Kassoer	*S. officinarum* × hybrids	Java, Indonesia	POJ2628, Mex73-206
Co453	B.Cheribon × Co285	*S. officinarum* × hybrids	Coimbatore, India	Co605, Co785, CR6368, CR67274, My5715, etc
CYC 71-370	Vietnam Niucane × Ya 58-47	*S. officinarum* × hybrids	Yacheng, Hainan	CGN 81-1035
Co223	Chittan × M1515	*S. officinarum* × hybrids	Coimbatore, India	No
Comus	Oramboo × Q813	*S. officinarum* × hybrids	Australia	No
MQ28-13	Badila × Q813	*S. officinarum* × hybrids	Australia	No
CYC 65-621	Badila × Kassoer	*S. officinarum* × hybrids	Yacheng, Hainan, China	No
CYC 97-27	Korp × CYC 96-68	*S. officinarum* × hybrids	Yacheng, Hainan, China	No
CYC 97-38, CYC 97-40	50Uahiapele × CYC 95-41	*S. officinarum* × hybrids	Yacheng, Hainan, China	No
CYC 98-11	Bailou cane × CYC 95-41	*S. officinarum* × hybrids	Yacheng, Hainan, China	No
CYR 88-196	Waie × Yunge F1189	*S. officinarum* × hybrids	Ruili, Yunnan, China	No
POJ2747	POJ2628 × Lahaina	hybrids × *S. officinarum*	Java, Indonesia	POJ2927, POJ2928, POJ2929
CYC 59-818	S17 × Badila	hybrids × *S. officinarum*	Yacheng, Hainan, China	CYZ 71-388, CLH 78-337
Damon	Trojan × Oramboo	hybrids × *S. officinarum*	Australia	No
PT40-196	POJ2725 × Badila	hybrids × *S. officinarum*	Pingtung, Taiwan, China	No

Source: Wu et al., 2014.

4.1.5 Cross between *S. officinarum* and related wild species

In recent years, many *S. officinarum* and related wild species have been hybridized in China. There are three reported hybridization models: *S. officinarum* × *Erianthus rufipilus*, *S. officinarum* × *Erianthus rockii,* and *S. officinarum* × *Erianthus arundinaceus*, and each method obtained seeds and seedlings. Among them, *S. officinarum* × *Erianthus rufipilus* obtained seeds from one cross combination; In the way of *S. officinarum* × *Erianthus rockii*, fuzzy was produced from 10 cross combinations, 9 *S. officinarum* and 4 *Erianthus rockii*. Among the 10 cross combinations, the highest number of seedlings reached 178. And several drought tolerant and highly rust-resistant progenies were obtained through further backcrossing; The hybridization of *S. officinarum* × *Erianthus arundinaceus* has been carried out earlier, and fuzzy from 28 cross combinations obtained (Table 3-5), not only F_1 progenies were generated, but also BC_{2-4} were developed, those progenies performed well drought and freeze tolerance, it is expected that the cultivar with the genetic background of *Erianthus arundinaceus* will be developed (Table 3-4, table 3-5 and table 3-6).

4.1.6 Interspecific hybridization between *S. officinarum* and wild species within the genus

Crossing *S. officinarum* to the wild relatives within *Saccharum* L. is easy to succeed. Crossing *S. officinarum* with *S. spontaneum* and with *S. robustum* are major approaches. To date, the hybridization mode between *S. officinarum* and *S. spontaneum* is merely under *S. officinarum*× *S. spontaneum* (Table 3-9 and table 3-10), which is the most effective model. Due to the successful input of the genetic background of *S. spontaneum* into modern cultivars, the growing area of sugarcane has been continuously expanded, and breakthroughs have been made in planting latitude and altitude, thus promoting the continuous development of the cane sugar industry. The hybridization between *S. officinarum* and *S. robustum* can improve the growth vigor of sugarcane, increase the hardness of sugarcane stalk, have high fiber content, and greatly enhance the wind tolerance, drought tolerance, and ratooning ability of obtained progenies. There are two ways of cross utilization between *S. officinarum* and *S. robustum*s (Table 3-12). Although there has been no report of succeeding in developing a cultivar so far, beneficial exploring on its utilization in hybridization has been conducted. With the development of science and technology and productivity, sugarcane production accelerates the transition from artificial to mechanized. With the increasing importance of lodging tolerance, the hybridization scale between *S. officinarum* and its wild relatives may increase. The results show that:

Firstly, *S. officinarum* × *S. officinarum* and *S. officinarum* × *S. spontaneum* are the earliest hybridization ways, the most hybrids, and the most outstanding breeding achievements. At present, all the well-known varieties are from these two ways. More than 700 *S. officinarum* in the world, but merely about 10 of them have been successfully used for sugarcane breeding. The number of *S. spontaneum* is greater than *S. officinarum*; however, most of them have not been well utilized in sugarcane genetic improvement. It is expected to make a more significant contribution to sugarcane breeding by exploring and hybridizing new *S. officinarum* with excellent characters and *S. spontaneum* to carry out basic hybridization. Further hybridization within equal generations may contribute more to sugarcane breeding.

Secondly, according to the flowering characters of *S. sinense*, landrace, and *S. barberi*, we believe that *S. sinense* and *S. barberi,* which are accessible to blossom, are the hybrids of *S. officinarum* and *S. spontaneum*. In contrast, those which are difficult to blossom are the hybrids of *S. officinarum* and *Miscanthus sinensis*. There are two kinds of views because the research objects of Luo and Grassl are different. According to this viewpoint, it is easy to explain why only a few *S. sinense* (Uba), *S. barberi* Chunnee, and Kansar have contributed to sugarcane breeding, while other *S. sinense*, landrace, and *S. barberi* have poor breeding performance.

4.2 The Hybrid Utilization Mode and Effect of *S.barberi*

4.2.1 Interspecific hybridization between *S. barberi* and *S. officinarum*

There are four types of *S. barberi* (Wang, 1979):*Sunnabile, Mungo, Nargori, and Saretha*. Among them, Chunnee and Kansar are the most widely hybridized in *Saretha* type. Hybrid utilization between *S. barberi* and *S. officinarum* is early, and the number of hybridization was low. Chunnee had the most significant breeding performance that has developed many essential parents (Table 4-4) and has been used to develop many varieties (described in 4.3.1.3 for details). However, the main focus of modern sugarcane breeding has been on the research and utilization of the original parents, but the hybrid between new *S. officinarum* and *S. barberi* has rarely been carried out. The main problems in hybrid utilization of other *S. barberi* are flowering, poor pollen development, a small number of progenies, etc.

4.2.2 Hybridization between *S. barberi* and hybrids

The hybridization between *S. barberi* and hybrids includes *S. barberi* × hybrids and hybrids × *S. barberi*. Sugarcane breeders have carried out many fruitful interspecific hybridizations between *S. barberi* and *S. officinarum* and tried the hybridization between *S. barberi* and hybrid (Table 4-6). The main methods are *S. barberi* × hybrid and hybrid × *S. barberi*.

Table 4-6 Utilization of *S. barberi*

Cross combination	Hybridization type	Location	Varieties/clones	Derived varieties
B.Cheribon × Chunnee	*S. officinarum* × *S. barberi*	Java, Indonesia	POJ213, POJ143, POJ181, POJ228, POJ234, POJ369, POJ826, POJ920, POJ979, POJ2379	Many
G.Praenger × Chunnee	*S. officinarum* × *S. barberi*	Java, Indonesia	POJ26, POJ33, POJ36, POJ139	No reports were reported
ZW Cheribon × Chunnee	*S. officinarum* × *S. barberi*	Java, Indonesia	POJ1410	No reports were reported

				continued
Cross combination	Hybridization type	Location	Varieties/clones	Derived varieties
POJ213 × Kansar	Hybrids × S. barberi	Coimbatore, India	Co213	Many
Chunnee × POJ100	S. barberi × Hybrids	Java, Indonesia	POJ385	Many

Source: Wu et al., 2014.

4.2.2.1 S. barberi × hybrids

In Java, Indonesia, sugarcane cultivar POJ385 was developed through the hybridization between S. barberi and Chunnee, and it was further crossed to a hybrid POJ181, and a better cultivar POJ1499 was developed. In India, Co213 × POJ1499 developed Co301. Co301 was crossed to CP29-103 as a female (Co301×CP29-103), and the cultivar CL41-114 was developed in Florida in 1941; In India, P4383 and Co603 were crossed to Co301 as females and developed Co1148 and Co678. POJ385 and its derivatives were used as parents in Taiwan province of China and South Africa and developed popular cultivars F105, N10, and N55-805 that were widely grown.

4.2.2.2 Hybrid × S. barberi

The cross combination POJ213 × Kansar was hybridized in Coimbatore, India, and the sugarcane cultivar Co213 was developed. Its female parent, POJ213, was the hybrid of B.Cheribon and Chunnee, so it contained a 75% genetic background of S. barberi (Kansar and Chunnee accounted for 50% and 25%, respectively). Co213 had been widely grown worldwide, and it has been an essential parent.

4.3 The Way and Effect of Hybrid Utilization of S.sinense

There are about 38 genotypes of S. sinense, mainly including Uba, zhucane, and Lucane. zhucane is not easy to flower, and its pollen is sterile, so few reports are on its hybridization. Lucane is a landrace in Sichuan, China, weak in poor wind tolerance, easy lodging, low sugar content, low yield, no heading, no flowering. Uba's pollen is mainly sterile, and the seedlings produced have poor performance, so the cross-utilization of *S. sinense* in sugarcane breeding is rare. The main utilization modes are as follows:

4.3.1 Cross between *S. sinense*, landrace, and *S. officinarum*

At the end of the 20th century and the beginning of the 21st century, HSBS and RSBS also made significant breakthroughs in the hybrid utilization of new *S. sinense*, landrace, and *S. officinarum* (Table 4-3). Seedlings were obtained from a range of cross combinations. Clones were advanced from the progeny populations, which laid the foundation for further utilization (details are described in 4.3.1.2).

4.3.2 Crossing within *S. sinense*

There is a rare report on the utilization of *S. sinense* in sugarcane breeding, particularly the successful crossing within S. sinense. In recent years, RSBS started from April 19 to August 17 with a fixed photoperiod treatment of 12 h and 20 min, then reduced the light for 60 s every day for 20 days, and then fixed for 12 h every day, which induced a batch of *S. sinense* to blossom successfully, and successfully obtained fuzzy of *S. sinense* × *S. sinense* (Guangze bamboo cane × Henan Xuchang chewing cane), and obtained 4 seedlings. Three of them were true hybrid progenies (Tao Lian'an et al., 2011). The breeding effect of hybrid progenies still needs to be further verified.

4.3.3 Hybridization between *S. sinense* and hybrid

Cross-utilization between *S. sinense* and hybrid is an earlier and more effective utilization mode (Table 4-7). The hybridization of Uba in Hawaii and Barbados generated several cultivars. For example, cultivar H28-4399 was developed in Hawaii through the hybridization of Uba × H450; then, it was further used as the male parent to breed H32-1063 (POJ2878 × H28-4399). Moreover, the hybridization of Uba × D1135 was successfully carried out, and the cultivars Mo1929 and H33-6705 were developed through their further crossing and backcrossing in Hawaii. In Canal Point of Florida, the hybridization of POJ213 × H.Uba generated the cultivar

CP726, and further crossing and backcrossing generated a range of cultivars CP27-28, CP27-108, CP28-78, CP52-105, CP63-306, CP59-50, CYT63-237, CL47-143, CP33-30, CP36-819, F36-819, CP43-64, CYC66-58, CYC57-20, CYC66-58, CYN81-762, ROC22 and so on. The crossing of Uba generated cultivars B45151, B45258, and B54142 also in Barbados (Peng, 1990). HSBS successfully hybridized Tanzhouzhucane × F134+Co331, Vietnam Niucane × CYC 58-46, and Guangxi zhucane × CYC 58-63 and generated F_1 clones CYC 57-36, CYC 71-370, and CYC 73-92, but have no new cultivar developed yet.

Table 4-7 Hybridization between *S. sinense* and hybrids

Variety/Clone	Cross combination	Type	Location	Variety developed
CP726	POJ213 × H.Uba	Hybrid × *S. sinense*	Canal point. USA.	Many
UD50	Uba × D1135	*S. sinense* × hybrids	Hawaii, USA	Many
H28-4399	Uba × H450	*S. sinense* × hybrids	Hawaii, USA	Many
CYC 71-370	Vietnam Niucane × CYC 58-46	*S. sinense* × hybrids	Yacheng, Hainan, China	No
CYC 57-36	Tanzhou zhucane × F134 + Co331	*S. sinense* × hybrids	Yacheng, Hainan, China	No
CYC 73-92	Guangxi zhucane × CYC 58-63	*S. sinense* × hybrids	Yacheng, Hainan, China	No

Source: Wu et al., 2014.

4.3.4 Cross between *S. sinense* and related wild species

The hybridization between *S. sinense* and related wild species is mainly between *S. sinense* and *Erianthus arundinaceus*, but the hybridization between *S. sinense* and other related species has not been reported. The way of hybridization between *S. sinense* and *Erianthus arundinaceus* is *S. sinense*× *Erianthus arundinaceus*. Since the 1990s, HSBS occasionally crossed 10 combinations, and all the families excepted one have produced seedlings. Progenies have been obtained through the early hybridized combinations (Table 3-6); however, this hybridization has not been cultivated yet.

4.4 Ways and Effects of Interspecific Hybridization for Original Saccharum Cultivated Species

Two or three kinds of excellent genes have been aggregated in the progeny of *Saccharum* L. (*S. officinarum*, *S. barberi,* and *S. sinense*), and the genetic basis is more comprehensive. Different traits such as yield, sugar content, disease resistance, stress tolerance, and adaptability are improved compared with the original species. For example, the resistance to Xiangmaobing was better in POJ100 (Loethers × Black Bangjermasin Hitan) than Black Cheribon; POJ213 (Black Cheribon × Chunnee) performed better plant shape, better ratooning ability, and higher sugar content than its parents. Those mentioned above "noble" varieties of the first basic hybridization not only play a positive role in the development of the local sugar industry but also serve as essential parents in the late sugarcane breeding, which is repeatedly hybridized and backcrossed for utilization, and developed a large number of excellent varieties/clones with different genetic background, thus promoting the development of world sugar industry. Since the hybridization of sugarcane, Java in Indonesia and Hawaii in the United States have contributed to early sugarcane breeding. The first generation of excellent "noble" varieties developed using the original sugarcane species are D74, EK2, H27-8101, etc. At present, the sugarcane varieties popularized in production at home and abroad contain the above F_1 noble species' genetic background.

The primary deficiency is that there are more than 700 *S. officinarum*, and up to now, only 10 of them have been used in sugarcane breeding. Otaheiti, Black Cheribon, Kaludai Boothan, Black Bandjermasim Hitan, Ashy Mauritius, D74 (naturally pollinated offspring of Critalina), Badila, Lahaina, Yellow Caled, Batjam, Fiji, Vellai, S. Mauritius, etc. (Peng, 1990; Zhang et al., 2006); To develop more worldwide varieties, Wu Caiwen first put forward the concept of developing improved, innovative and independent parent system by using new original species in 2005, and put forward the method of developing breakthrough sugarcane varieties.

Over 100 years of breeding, the shortcomings of interspecific hybridization of sugarcane varieties are mainly manifested in the following three aspects: first, there are few successful examples of interspecific hybridization of new progenitor; second, in the recent hybridization of some progenitors, the old progenitors that have been used are still used, and the new progenitors are less used; third, the new independent parent system has not been produced because of the few materials that can be used for the further development of the peer hybridization.

References

AN R D, CHU L B, SUN Y F, et al., 2007. Breeding of new sugarcane variety Yunzhe 99-155[J]. Sugarcane Sugar Industry (3):7-15.

BRANDES E W, SARTORIS G B, 1936. Sugarcane: Its origin and improvement[R]. USDA. Year Book.

BUZACOTT J H, 1950. Recent trends in the production of cane seedlings in Queensland by the Bureau of Sugar Experiment Stations[J]. South African Sugar Journal (34): 721-727.

CHEN R K, LIN Y Q, ZHANG M Q, et al., 2003. Theory and practice of modern sugarcane breeding[M]. Beijing: China agriculture press (in Chinese).

CHEN X W, DENG H H, CHEN Y S, 2010. Utilization of Badila in the breeding of YC-Series Parents and new varieties of sugarcane[J]. Sugarcane and canesugar (6):1-5.

CHU L B, 2000. Study on Breeding System of "YN" Sugarcane —— Using "Heterogeneous Compound Separation Theory" to obtain new germplasm with superior high sugar content in Yunnan *S. spontaneum* F_1[J]. Sugarcane, 7 (4):22-33.

DENG H H, LI Q W, CHEN Z Y, 2004. Innovation and utilization of sugarcane parents[J]. Sugarcane, 11 (3): 7-12.

DENG H H, ZHOU Y H, XU Y I, et al., 1996. Analysis on the genetic relationships of the major sugarcane clones in China[J]. Sugarcane and Canesugarr (6):1-8.

D'HONT A, GRIVET, FELDMANN P, et al., 1996. Characterisation of the double genome structure of modern sugarcane cultivars (*Saccharum spp.*) by molecular cytogenetics[J]. Molecular and General Genetics (250):405-413.

FU C, LIU S M, YANG Y H, et al., 2004. Review on research and utilization of sugarcane germplasm in Hainan Sugarcane Breeding Station during the Ninth Five-Year Plan[J]. Sugarcane and Canesugar (3):1-8.

HUANG Q Y, 1991. Viewing the breeding ways of China and the contribution of Hainan Sugarcane Breeding Station from the new sugarcane varieties bred in 1980s[J]. Sugarcane and Canesugarr (4):7-14.

HUANG Q Y, 1983. Summary of sugarcane breeding procedures[J]. Sugarcane and Canesugarr-Sugarcane (1):8-14.

LI Y R, 2010. Modern sugarcane science[M]. Beijing: China Agricultural Press (in Chinese).

LIN R J, 1987. Genetic analysis of pedigree of self-bred sugarcane varieties in China[J]. Sugarcane Sugar Industry (7):10-18.

LIU H B, 1997. Research and utilization of sugarcane distant hybridization in mainland China[J]. Sugarcane, 4 (3):7-9.

LIU S M, WANG L N, HUANG Z X, et al., 2011. Utilization of sugarcane parents of Yacheng series in sugarcane breeding of China[J]. Sugarcane and Canesugarr (4):5-10.

LUO J S, 1984. sugarcane science[C]. Guangzhou: Guangdong sugarcane society.

PENG S G, 1990. Sugarcane Breeding[M]. Beijing: Agricultural Press (in Chinese).

PIPERIDIS G, D' HONT A, 2001. Chromosome composition analysis of various Saccharum interspecific hybrids by genomic in situ hybridization (GISH)[J]. International Society Sugar Cane Technologists (24):565–566.

SOBHAKUMARI V P. 2005. Chromosome numbers of some noble sugarcane clones[J]. Journal of Cytology and Genetics (6):25-29.

STEVENSON G C, 1965. Genetics and breeding of sugarcane[M]. London: Longmans, Green and Co.

WANG J M, 1985. Sugarcane Cultivation in China[M]. Beijing: Agricultural Press (in Chinese).

WANG L P, FAN Y H, CAI Q, et al., 2003. Research progress on hybrid utilization of sugarcane germplasm resources[J]. Sugarcane (3):7-23.

WANG Q Z, 1979. Sugarcane farming[M]. Taipei: National Editorial Center.

WEN Y, 1998. Intergeneric hybridization and chromosome behavior of sugarcane related plants[J]. Sugarcane and Canesugarr (3):1-7, 17.

WU C W, 2005. Discussion on germplasm innovation and breeding breakthrough varieties in sugarcane[J]. Southwest China Journal of Agricultural Sciences, 17 (6):858-861.

WU C W, PHILLIP J, LIU J Y, et al., 2011. Inheritance of quality traits of the distant crossing between *S. officinarum* and *S. spontaneum*[J]. Journal of plant genetic resources, 12 (1), 59-63.

WU C W, ZHAO P F, XIA H M, et al., 2014. Modern cross breeding and selection techniques in sugarcane[M]. BeiJing: Science Press (in Chinese).

XIAO F J, LI Q W, DAI Y B, 1995. Observations on the agronomic, milling and botanical characters and the chromosome in commercial cane × *E.arundinaceus* hybrid and its parents[J]. Journal of Yunnan agricultural university, 10 (1): 29-34.

ZHANG M Q, DENG Z H, CHEN R K, et al., 2006. Genetic improvement and efficient breeding of sugar crops[M]. Beijing: China Agricultural Press (in Chinese).

5 Cultivation Methods of Improvement, Innovation and Independent Parent System in Sugarcane Hybrid Breeding

Sugarcane is the main sugar crop in the world. In the early 1990s, global sugar production was about 118 million tons, of which cane sugar production accounted for 81 million tons. In 2011, global sugar production reached 172 million tons, while cane sugar production increased to 138 million tons. The proportion of cane sugar production in global sugar production had increased from 68% to over 80%, while the ratio of beet sugar decreased evidently during the last 20 years. Thus, an excellent cultivar directly related to cane sugar production is the vital basis of sugar production. At present, most worldwide sugarcane varieties are bred by traditional hybridization, which could be divided into the basic cross and productive cross. Basis hybridization aims to create improved, innovative, or breakthrough parent and productive hybridization to cultivate superior varieties by utilizing these hybrid parents.

Basic hybridization is the origin, foundation, and prerequisite of plant crossbreeding. Sugarcane breeders usually need to perform multiple hybridization rounds to generate excellent parents by varied hybrid strategies, such as pedigree selection, backcross breeding, etc. The creations of new superior parents are complex and time-consuming but are also essential to breed new cultivars. The scale of hybrid strategy had been universally recognized, and their effects of hybrid utilization were discussed. Many basic germplasms, such as *S. officinarum*, *S. spontaneum*, *S. barberi*, *S. sinense*, *S. robustum*, and *S. edule*, were used to make basic hybrids in the century sugarcane breeding program (see Chapter 4 for details). It was reported that there were 33 ways of basic hybridization and 258 hybrid combinations, of which 11 ways of basic crosses and 98 relative combinations had made outstanding contributions to sugarcane breeding (Table 5-1). It was evaluated that 40 hybrid combinations were effective, which led to breeding varieties successfully. The success ratio of hybrid combinations was up to 42.5% of all reported hybrid types. Most offspring of basic crossings contained a partial mixture of *S. officinarum*, *S. spontaneum*, *S. barberi*, *S. sinense, S. spontaneum,* and *S. robustum*. The phenomena were also found in recent reports. The successful methods and number of crosses used in sugarcane breeding programs will continue to increase over time. Diverse ways of hybridization usually make different contributions to sugarcane breeding. According to the basic cross in sugarcane breeding program types and contributions, hybrid parents can be divided into independent parents, innovative parents, and improved parents.

Table 5-1 Classification of the important basic cross in sugarcane breeding

No.	Hybrid type	Times of cross	Number of cultivars
1	*S. officinarum*×*S. officinarum*	11	5
2	*S. officinarum*×*S. barberi*	3	1
3	*S. officinarum*×*S. spontaneum*	20	10
4	*S. officinarum*×hybrids	13	5
5	*S. sinense*×hybrids	6	2
6	*S. barberi*×hybrids	1	1
7	hybrids×*S. officinarum*	4	2
8	hybrids×*S. robustum*	9	3
9	hybrids×*S. spontaneum*	29	4
10	hybrids×*S. barberi*	1	1
11	hybrids×*S. sinense*	1	1

Source: Wu et al., 2014.

5.1 Independent Parents Breeding and Its Contribution to Sugarcane Breeding

Hybridization is a method to breed varieties by creating new variations through cross-breeding among different materials. It is also the most commonly used, most popular, and most effective method at present. Its significance of crossbreeding lies based on heterosis. The theoretical foundation of heterosis belongs to the gene recombination of the hybrid progeny. One basic principle of exploiting heterosis is to select good hybrid combinations. There is higher heterosis when the parents show distant genetic relationships, significant differences in characters, and high heterogeneity. Genetic variation and basis are important restrictions in breeding sugarcane varieties. Creating a sugarcane independent parent system aims to screen out superior *Sacchuarum* species by identifying and evaluating enormous germplasm resources, including *S. officinarum*, *S. sinense*, *S. barberi*, *S. spontaneum*, and *S. robustum*. The materials with fine characters are hybridized as parents to produce the first filial generation (F_1, the first noble cane). Excellent F_1 progeny is selected to make a peer-to-peer cross with the F_1 of another independent hybridization. The F_2 progeny (the second noble cane) contains an equal consanguinity proportion from 4 original ancestors. The potentiality of heterosis could be substantial, making it easy to select the progeny with breakthrough traits in yield, sugar content, ratooning, and stress resistance. At present, all worldwide breakthrough sugarcane varieties or parents, such as POJ2878, Co290, CP49-50, and Co419, were successfully bred by this method. The parent system, consisting of four and more different *Saccharum* species and generated by peer-based hybridization, was named the independent parent system by Wu et al. (2005).

5.1.1 The Independent Parent System with Four Original *Saccharum* Species

The independent parent system with four original *Saccharum* species was used for breeding sugarcane varieties in the century of the sugarcane breeding program. Many elite varieties or parents were bred within an independent parent system, such as EK28, POJ2364, Co213, Co221, Co244, Co281, etc. These varieties or parents are the most successful models of cane nobilization. The hybrid modes of original *Saccharum* species mainly include *S. officinarum* × *S. officinarum*, *S. officinarum* × *S. spontaneum,* and *S. officinarum* × *S. barberi*. The sugarcane varieties bred by the independent parent system have made outstanding contributions to developing the world's sugarcane industry. Unfortunately, there is no report that the progeny of *S. officinarum* × *S. robustum* is widely used in sugarcane breeding.

5.1.1.1 EK28

EK28 was bred in Java Sugarcane Breeding Station, Indonesia, which has created the most successful precedent of tropical intraspecific hybridization. The four *Saccharum* species all belong to *S. officinarum* (Figure 5-1). Before interspecific cross-breeding, EK28 contributed to the Java cane sugar industry during the transitional period (Luo, 1984). Its central growing region was limited in Java, Indonesia, but it was difficult for EK28 to spread to other areas. Despite its outstanding contribution to the Java sugar industry, EK28 could not plant in other sugarcane areas. The main reason for its limited promotion might lie in its kinship, which was the only origin from *S. officinarum*. *S. officinarum* has the advantages of high yield, high sugar, big stem, and low fiber content and has the disadvantages of disease resistance, stress resistance, and adaptability. Due to consanguinity defects, EK28 could not be widely spread.

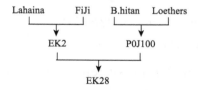

Figure 5-1 The pedigree chart of EK28

5.1.1.2 POJ2364

POJ2364 was also bred in Java Sugarcane Breeding Station, Indonesia. It is one of the best parents bred by crossing *S. officinarum* with *S. spontaneum*. The original species of POJ2364 have 3 *S. officinarum* and 1 *S. spontaneum* (Figure 5-2). Due to containing the consanguinity of wild species, POJ2364 shows significantly improved stress resistance and adaptability.

5 Cultivation Methods of Improvement, Innovation and Independent Parent System in Sugarcane Hybrid Breeding

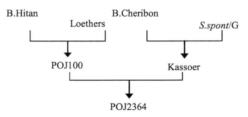

Figure 5-2 The pedigree chart of POJ2364

5.1.1.3 Co221, CO244, and Co281

Co221, CO244, and Co281 were bred in Coimbatore SugarcaneBreeding Station, India (Figure 5-3, Figure 5-4, and Figure 5-5). Their parents are origin from *S. officinarum, S. spontaneum,* and *S. barberi*. Although the four original species of the three varieties are not the same germplasm, their parents consist of 2 *S. officinarum*, 1 *S. barberi,* and 1 *S. spontaneum*.

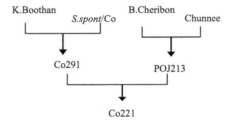

Figure 5-3 The pedigree chart of Co221

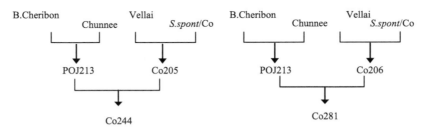

Figure 5-4 The pedigree chart of Co244   Figure 5-5 The pedigree chart of Co281

The independent parent system with four *Saccharum* species has different hybrid types and original ancestors, but the common feature is that the pedigree chart is an inverted triangle. The consanguinity proportion of each *Saccharum* species is equal to 25%. These parents are all economically cultivated at different periods in different sugarcane planting regions. But these parents either have a high consanguinity proportion of *S. officinarum* (75% in EK28) and insufficient adaptability or have a higher consanguinity proportion of *S. spontaneum* (25% in POJ2364, Co221, Co244, and Co281), and the commercial traits were poor. Their agricultural characters are not perfect. Although they are planted in different production areas, the promotion areas are not significant. Because of their excellent consanguinity, they have become the most critical parents of noble cane breeding in the world. At present, all sugarcane varieties in the world contain the kinship of these varieties. Through continual mutual crossing or back-crossing of them and their offspring, many worldwide varieties and parents have been bred, making significant contributions to developing the global sugarcane breeding and sugar industry. Since the 1930s, due to the focus on the existing fundamental hybridization, no new parent system similar to the above genetic relationship has been produced, resulting in few breakthroughs in sugarcane breeding.

5.1.2 Breakthrough Sugarcane Varieties and Parents Containing 6 or More Original *Saccharum* Species

Since the Dutch breeder Jeswiet (1890) proposed the Theory of Sugarcane Noblization Breeding, breakthrough sugarcane varieties and parents had been bred and recognized worldwide, including POJ2878, Co290, Co419, F134, Nco310 and CP49-50, and so on. These varieties have a common characteristic. Their basic crosses are derived from 3 types of crossbreeding containing *S. officinarum* × *S. officinarum*, *S. officinarum* × *S. barberi,* and *S. officinarum* × *S. spontaneum*, all of which are independent parents of those mentioned above four original species. The hybrid offspring, produced by peer-based parental crossing, have significant heterogeneity, distant genetic relationship, and prominent heterosis (Wu et al., 2005). They are widely used in production and used as hybrid parents to directly breed many excellent varieties.

5.1.2.1 POJ2878

POJ2878 contains 6 original cultivated species (Figure 5-6). It is the third-generation noble cane bred by Jeswei at Java Sugarcane Breeding Station in 1921. It shows a large stem diameter and high sugar content like the parent B. Cheribon. This variety exhibits cyan epidermis, resistant to disease, and adaptable similar to the parent *S. spontaneum* in *Glagah*. It presents lovely crop architecture and yields. It grows well in both tropical and subtropical sugarcane regions, showing wide adaptability. In 1929, the promotion area accounted for 95% of the total sugarcane planting areas in Java and spread rapidly, proving that POJ2878 was loved by sugarcane farmers and sugar mills in Java. In 1930, its reputation spread all over the world. It was introduced to sugar mills as raw materials and used as parents for many breeding organizations. At present, almost all varieties in the world contain the consanguinity of POJ2878, which was recognized as the first generation of 'Sugarcane King'. However, this variety is susceptible to Fiji disease and downy mildew (Luo, 1984). From the kinship perspective, the primary defect of hybridization type is that 2 original cultivated parents (B.hitan and Loethers) repeatedly used twice, leading to a small amount crossover of the kinship (25%). Suppose one of POJ100 was replaced with another F_1 germplasm, which parents cannot be the same as the existing basic germplasm in the system. In that case, the value of POJ2878 may be more prominent in sugarcane production and hybridization utilization.

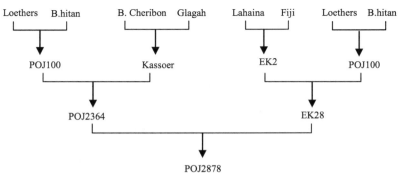

Figure 5-6 The pedigree chart of POJ2878

5.1.2.2 Co290

Co290 contained 6 original ancestors (Figure 5-7) and was bred at Coimbatar SugarcaneBreeding Station in India. It is a hybrid offspring of *S. officinarum*, *S. barberi,* and *S. spontaneum*. The variety has a thick, soft, and short stem, red skin, compact plant shape, and ratooning solid ability but is susceptible to mosaic disease and red rot. This variety is an economic cultivar in India, Argentina, and Louisiana. In 1937, Co290 was introduced to the Neijiang Sugarcane Experiment Stations, Sichuan, China, from the Agriculture Department of the United States. The yield of this variety was better than that of local sugarcane varieties. In the 1950s, it replaced local varieties and became the primary economic cultivar. This variety has good pollen fertility and can be used as a female parent or used as a male parent. Many excellent varieties and parents were bred by directly using Co290 as parents in many sugarcane breeding units worldwide.

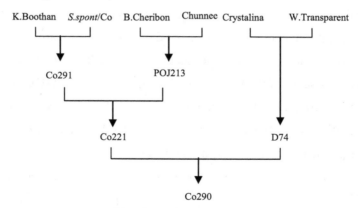

Figure 5-7 The pedigree chart of Co290

The current varieties almost contain the consanguinity of Co290 (Luo et al., 1984; Peng et al., 1990), which reputation ranks first with POJ2878. From the pedigree chart perspective, there is no crossed consanguinity among the parents of the variety. There is no consanguinity crossing between the original parents of Co290. The primary defect is that Co221 and D74 are unstructured peer-to-peer crossing, resulting in a large proportion of kinship between the two *S. officinarum* of D74, which makes Co290 and its offspring belong to soft cane, with soft sugarcane skin, easy to lodging, severe rat damage, making significant production loss (Luo et al., 1984). If D74 and Co221 belong to peer-to-peer crossing, It is possible that Co290 has better traits, and the promotion area will be more widely. Using the supposed variety as the hybrid parent probably overcomes the shortcomings of softness. There will be a more excellent value of hybridization and better offspring.

5.1.2.3 Co419

Co419 has an excellent genetic basis containing 12 original *Saccharum* species. It is known worldwide as the 'Sugarcane King'. Its parental combination is POJ2878×Co290, which parents are both world-renowned hybrid parents. Since the development of sugarcane sexual crossbreeding in the mainland of China, 30 of the 241 fine sugarcane varieties are the direct hybrids of Co419. Co419 is the interspecific hybrid offspring of three species, including *S. officinarum*, *S. barberi*, and *S. spontaneum*. The consanguinity proportion of *S. officinarum*, *S. barberi*, and *S. spontaneum* is almost optimum allocation with 81.25%, 6.25%, and 12.5%, respectively. The odds of India *S. spontaneum* and the Java *S. spontaneum* are about even. By analyzing 12 original parents, three *S. officinarum* (B.Hitan, Loethers, and B.Cheribon) are repeated twice; the heterogeneous rate of parental genetics was 72.7%. The consanguinity proportion of *S. officinarum* is optimal, but there is a small amount of genetic crossover within *S. officinarum*. It is considered that a higher rate of parental heterogeneity usually shows a better genetic basis and breeding effect. Besides, the kinship of CHN 53-63 is identical to Co419. CHN 53-63 is also one of China's top ten sugarcane parents, contributing to China's sugarcane industry.

5.1.2.4 F134

F134 was developed by Sugarcane Industry Research Institute in Taiwan province of China. Its parent combination is Co290×POJ2878, which has similar parents as Co419, except that the male parent and the female parent are opposite (Figure 5-8). F134 shows good production performance, strong stress resistance, and wide adaptability. It used to be the primary cultivar in the main sugarcane planting regions of China. At the same time, this variety has good flowering and strong crossing affinity making it an excellent parent. Since the development of sugarcane crossbreeding in the mainland of China, more than 300 excellent varieties have been bred, 42 of which are the offspring of F134 (about 18% of the total bred varieties).

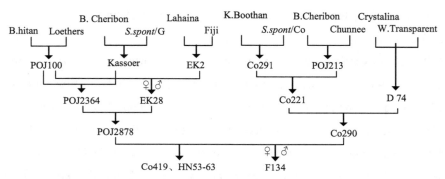

Figure 5-8 The pedigree chart of Co419, CHN53-63 and F134

5.1.2.5 Nco310

The hybridization fuzz of Nco310 was made in Coimbatore, India, and was bred in Natal. The parental combination is Co421×Co312. The consanguinity relationship is more complicated than that of Co419 and F134 (Figure 5-9). The type of hybrid is different and lacked the symmetry of crossing. Co285, the F_1 generation of *S. officinarum* and *S. spontaneum*, is directly crossed with POJ2878, which belonged to the F_3 generation. Their offspring Co421 thus returns to the F_2 generation. The Kansar (belong to *S. barberi*) is crossed with the F_1 generation POJ213, and the offspring Co213 is returned as the F_1 generation of *S. barberi*. The two F_2 hybrids, Co421 and Co312, are finally crossed. Nco310 is the F_3 generation of *S. officinarum*, *S. barberi,* and *S. spontaneum*. This cultivar contains 11 original parents, including 7 *S. officinarum* (*S. mauritius*, B.Cheribon, B.hitan, Loethers, Vellai, Lahaina, Fiji), 2 *S. barberi* (Kansar and Chunnee), and 2 *S. spontaneum*, accounting for 53.125%, 25.0%, and 21.875%, respectively. Due to the low generations of *S. barberi* and *S. spontaneum*, it is highly resistant to stress, especially cold and disease. It is suitable for planting in subtropical and temperate regions. It is widely planted in Taiwan, Sichuan, Jiangxi, and other provinces of China and has made unique production and breeding contributions. Due to a small amount of confusion in the consanguinity relationship, its role as a hybrid parent is not as outstanding as F134 and Co419. Since sugarcane crossbreeding in the mainland of China, 15 excellent sugarcane varieties are directly bred using Nco310 as a parent.

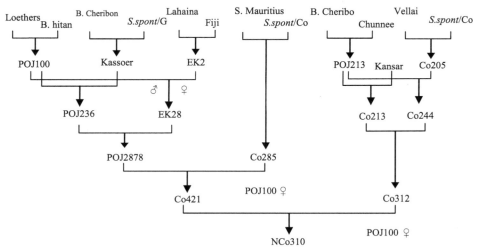

Figure 5-9 The pedigree chart of NCo310

5.1.2.6 CP49-50

CP49-50 was bred at Canal Point in the United States. The parental combination is CP34-120×Co356, and the kinship is relatively straightforward (Figure 5-10). This variety contains 9 original *Saccharum species*, including 6 *S. officinarum* (Vellai, B.Cheribon, B.hitan, Loethers, Lahaina, Fiji), 1 *S. barberi* (Chunnee), 2 *S. spontaneum* (came from India and Java). The proportions of *S. officinarum, S. barberi*, and *S. spontaneum* were 80.75%, 6.25%, and 13.0%, respectively. Look at the family relationship; Co281 is the F_2 generation of *S. officinarum, S. barberi*, and *S. spontaneum*. It was directly used to cross with F_3 hybrid POJ2878, and the offspring CP34-120 was back to the F_3 generation. From the perspective of kinship, Co356 was the hybrid offspring of POJ2725 and *Sorghum*. But according to Deng et al. (2004), Co356 did not contain the correlative of *Sorghum*, indicating that the distant hybridization was unsuccessful. Co356 was the offspring of POJ2725 by self-crossing or internal crossing, which was still the F_3 generation. Finally, the F_4 generation was bred by the cross of two F_3 generations. The genetic symmetry of CP49-50 was increased, which made its traits more excellent. This variety exhibits erect phenotype, tillering, strong ratooning, wide adaptability, and resistance to disease, wind, water, drought, cold, and floods. It also shows excellent yield and sugar content and is suitable to act as a hybrid parent. According to statistical results, up to 42 of 300 excellent sugarcane varieties are the direct hybrid progeny of CP49-50 in the mainland of China.

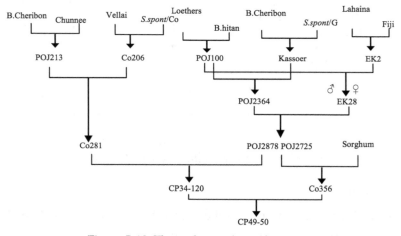

Figure 5-10 The pedigree chart of CP49-50

5.1.2.7 CGT11

CGT11 was bred in Sugarcane Research Institute, Guangxi Academy of Agricultural Sciences (GXSRI). Its parental combination is CP49-50×Co419, showing a clear genetic relationship but higher hybrid generations (Figure 5-11). This variety contains 12 original *Saccharum* species, among which 9 *S. officinarum*, 1 *S. barberi* (Chunnee), and 2 *S. spontaneum* (one from India and the other from Java). The consanguinity proportions of *S. officinarum*, *S. barberi* (Chunnee), and *S. spontaneum* were accounting for 80.75%, 6.25%, and 13.0%, respectively. This cultivar has many advantages, such as good tillering, early maturing, high yield, high sugar content, drought resistance, wide adaptability, ratooning solid, high stem-forming rate, and many stalks. It has been widely used as the leading cultivar in Chinese sugarcane growing regions. The main problem of hybrid type is the lack of symmetry and a small amount of genetic crossing. Its parents are both worldwide varieties. It is harder to find a pair of parents with symmetrical generations, no crossover, or few crossovers among existing parents or varieties. It may be why this variety performs well in production but does not breed well as the parents. The same hybrid combination also produces CYT65-906 in Guangdong. Besides, similar varieties (include CYZ81-173, CST66-166, CMT65-16, CGT64-73, CGN64-151, CGN66-241, CGN65-484, CYT65-1378, and CYZ68-154, etc.) which were bred by the combination of Co419×CP49-50 had been a large promotion area in production, which are the same parents of CGT11 with opposite hybrid mode. The above varieties all have some promotion areas in production but are far less than CGT11. The hybridization between them is all sibling inbreeding, failing to breed excellent offspring.

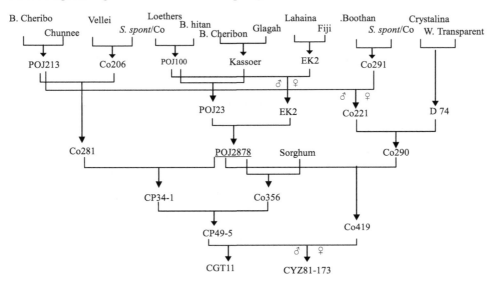

Figure 5-11 The pedigree chart of GT 11 and CYZ 81-173

5.1.3 Cultivation and Breeding Effect of Sugarcane Independent Parent System

The varieties cultivated by the independent parent system are abundant in quantity and good in trait quality. As hybrid parents, many varieties can be bred, spread in a large production area, making an outstanding contribution to the industry. From the kinship, the independent parent system's main characteristics are as follows: Firstly, basic hybridization is the interspecific hybridization. Secondly, the hybrid generation of each parent is the same. Thirdly, the consanguinity proportion of every original *Saccharum* is the same. Forthly, every hybridization combination is peer-based crossed. The more numbers of the original *Saccharum* species used and the more times of hybrid taken could produce the offspring's better traits performance. The less the genetic crossover between the original *Saccharum* species is, the stronger the bred varieties' breakthrough.

5.2 Cultivation of Innovative Sugarcane Parents and Their Contribution to Sugarcane Breeding

Parents' innovation improves some bad traits of existing parents by crossing new parents (2 or more). Since the second half of the 20th century, sugarcane breeders have done many sugarcane parents' innovations and bred many innovative parents and varieties. The hybrid method of the innovative parents is different from that of the independent parent system. The basic hybrid is mainly the unequal cross between the original species F_1 and non-F_1 hybrids and the method of breeding parents or varieties of sugarcane. The dendrogram mainly exhibits the asymmetry of hybridization and backcross between the original species F_1 hybrid with the higher generations such as F_2, BC_1, BC_2, or BC_3. The F_1 hybrid generated by the original *Saccharum* species is backcrossed with a higher generation hybrid from the consanguinity proportion. Although it does not make many back-crossing times, there are a more consanguinity proportion of old original species than that of new original species in the higher generation of hybrid. From the offspring's genotype, the genetic basis of the varieties has not changed much due to introducing only a few new original species.

Due to the limitation of genetic basis, there is no significant breakthrough in the characteristics of varieties. The production performance is not as good as that of the varieties produced by the independent parent system. Still, the number of varieties bred by this parenting system is not as many as that of the independent parent system. However, the breeding effect is better than that of improved parents due to introducing more new consanguinity. The representative germplasm included Co214, SW11, Uba Marot, H109, CYC58-47, CYC58-43, etc. Although they are excellent parents, they are not the most outstanding varieties all the time. Their offspring were limited in traits and promotion areas. These varieties can't become the 'Sugarcane King'.

5.2.1 The Effect of Direct Hybridization Between Original F_1 Hybrids with F_3 (or BC_2) Hybrids

Uba Marot, H109, SW11, and Co285 are all elite F_1 hybrid of original germplasm and are directly hybridized with F_3 hybrid (POJ2878). The difference is that Uba Marot and Co285 are F_1 hybrids of *S. officinarum* and *S. spontaneum* (Pan et al., 1990), while H109 (Lahaina × US1494) and SW11 (B.cheribon × Batjan) are the hybrid offspring of two *S. officinarum*. Although the original *Saccharum* species used are different, their F_1 hybrids (used as the paternal parent) are crossed with POJ2878 in Venezuela, Australia, Java, and India. Many bred varieties, such as V114, Q27, POJ2940, and Co617 (Figure 5-12). Meanwhile, a series of varieties were continued to hybridize to select new excellent progeny. For example, the offspring of V114 included R366, R397, R526, R541, R565, R566, and R567. The progeny of Q27 consisted of H31-1389, H40-1179, Q57, Q67, Q118, Q139, and Q80. The progeny of POJ2940 contained POJ2927, POJ2947, CB47-89, VMC71-238, PS8, and PS35, and the progeny of Co617 included Co853, CoS561, Mex60-1403, and CCZ4, etc. The pedigree and hybridization type of these varieties were similar, and their breeding effect was also similar. It indicated a strong correlation between cross-type and breeding effect. From the kinship perspective, the hybrids of the original *Saccharum* (such as SW11, Uba Marot, H109, and Co285) did not match any other F_1 hybrids. Still, they were crossed with the high generation hybrid POJ2878 (F_3), resulting in the asymmetry of hybridization and the lack of new *Saccharum* species. Therefore, it is reasonable to expect no breakthrough varieties and parents in the offspring of these innovative sugarcane parents.

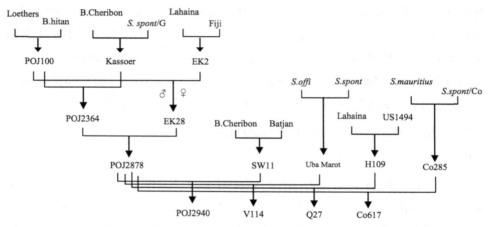

Figure 5-12 The pedigree chart of POJ2940, V114, Q27 and Co617 (F3×F1)

Similar parents also have CYC58-43, the F_1 germplasm of Badila, and Yacheng *S. spontaneum*. CYC 58-43 was crossed with F134 (BC_2 of *S. officinarum*), and the offspring were selected out of CYC62-40 (Figure 5-13). Due to the same hybridization way, the breeding variety has similar traits. However, the breakthrough in the characters is small, and the promotion area is limited.

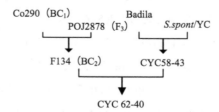

Figure 5-13 The pedigree chart of CYC62-40

5.2.2 Effect of Cross-Breeding between the F_1 Hybrid of the Original *Sacchuram* with BC_3 or Higher Generation

CYC 58-47 was bred by crossing Yacheng *S. spontaneum* with *S. officinarum* in the Hainan Sugarcane Breeding Station (HSBS). The F_1 hybrid was then backcrossed with CYT54-143, BC_3 generation of *S. officinarum,* and *S. spontaneum.* At last, innovative parent CYC71-374 was bred. There is no genetic crossover between the relatives of the dendrogram. It exhibits excellent hybrid symmetry(Figure 5-14). Ten sugarcane varieties that had accessed the approval or identification in China were directly bred using CYC 71-374 as a parent, including CCT89-103, CCT 89-241, CDZ 93-88, CDZ 93-94, CGN 95-108, CGT93-103, CGT94-119, CGT96-143, CGT96-44 and CYN 89-759. Their varieties were also used for breeding many excellent offspring, including CTC18, CGT27, CYZ89-7, and CYZ 99-91. These cultivars have the traits of *S. officinarum* (such as high yield, high sugar content) and specific adaptability. However, it only performs well under specific conditions for a particular cultivar, while the promotion scope and area are limited in other sugarcane localization. In terms of consanguinity relationship, there is no breakthrough in CYC71-374, although its parent, CYT54-143, is the offspring of 'Sugarcane King' POJ2878. CYC58-47, the F_1 hybrid of Yacheng *S. spontaneum* and *S. officinarum,* was not peer-to-peer crossed with other F_1 hybrids but a higher generation (BC_3), resulting in asymmetric consanguinity. Similar varieties included CYC79-290 and CYC 82-108. Despite the similar hybridization mode and breeding effect, the results of direct hybridization utilization were different. Due to different parents, the traits of the offspring were varied to some extent. For instance, the offspring of CYC79-290 were screened out to CYC84-153. This variety bred CGT94-116 by further hybridization with CGT71-5 (BC_6), also bred CYZ98-46 by hybridization with Co419 (BC_2). F16 and CGT2-761 are screened out by crossing CYC 82-108 with CP72-1210 (BC_4) and CYT 85-177 (BC_5). Due to few breakthroughs of consanguinity, these varieties are not prominent in traits, and their extension area is small.

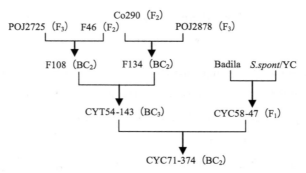

Figure 5-14 The pedigree chart of CYC71-374

5.2.3 Different Hybridization Utilization and Breeding Effect of F_1 Hybrid of the Same Original *Saccharum*

Different crossways of the F_1 hybrid of the same original species resulted in different breeding effects. CYC 58-43 is a F_1 hybrid of Badila (*S. officinarum*) and CYC *S. spontaneum*. It was continued hybridized with F134 (BC_2) to select CYC 62-40. By further hybridization and backcross utilization of CYC 62-40, three varieties were bred, including zhucane75-53, CGT89-5, and CFN 79-233. However, CYC 72-210 and CYT 64-395 were bred by the hybridization of CYC 58-43 with CYC 58-63 (BC_3) and HN 56-21 (BC_4), respectively. It indicated that the innovative F_1 parents of the same original germplasm could asymmetrically cross with different generations to produce varied varieties. However, the closer the generation is and the larger the consanguinity of the new original *Saccharum* species, the better the breeding effect and the more varieties breed (Figure 5-15).

The hybridization utilization of the innovative parental and their breeding effect show that the F_1 hybrid of the original species could be used to cross and backcross with the high generation. However, it is no better than the symmetric cross of the F_1 hybrid inbreeding effects, bred variety efficiency, and trait performance. In the same type of hybridization, the better the symmetric consanguinity is, the easier to breed varieties. In different hybridization modes, the closer the hybrid generation and the greater the kinship of the new original germplasm is, the better the breeding effect is. The hybrid effect of the original F_1 hybrid with F_3 (or BC_2) generation is better than that of the original F_1 hybrid with BC_3 or higher generation. CYC62-40 was bred by crossing CYC58-43 (F_1) with F134 (BC_2). More varieties were bred than crossing CYC 58-43 directly with BC_3 generation and BC_4 generation.

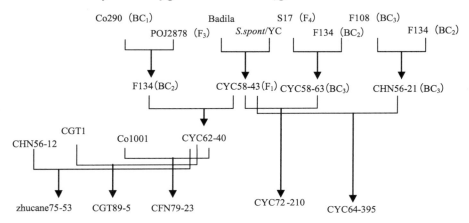

Figure 5-15 The pedigree of CYC58-43 and its descendants

5.3 Breeding Improved Sugarcane Parents and Their Contribution to Sugarcane Breeding

The improvement of parents is to improve few undesirable traits of existing parents by introducing few new genes (Wu et al., 2005). Since the second half of the 20th century, sugarcane breeders have done a lot of work to improve sugarcane parents, and many improved parents and varieties have been bred. The hybrid methods of breeding the improved varieties are different from those of the independent parent system and the innovative parent system. The basic hybridization methods of breeding the improved parents show that the original *Saccharum* species is directly crossed with the hybrids. Then the sugarcane parents or varieties are bred through asymmetric hybridization and backcross between the hybrids. It can be seen from the pedigree chart that new genetic germplasm is introduced by asymmetry hybridization. The continued hybridization is also in an asymmetric manner. The original *Saccharum* species (no matter *S. officinarum*, *S. barberi*, *S. sinense*, and *S. spontaneum*) directly hybridize with F_1, BC_1, BC_2 higher hybrids which makes the pedigree chart crucial asymmetry. From the proportion of consanguinity, the hybrid offspring are usually continued to backcross with the higher generations. Although the rounds of backcross are not frequent, the number of old germplasm is far beyond new germplasm. Because only a few new original species are used, the varieties' genetic bases have not changed. So the traits of the offspring also changed little. Due to the limitation of kinship, there is a slight variation in the characters of the offspring. These production performance and promotion areas are inferior to the independent parent system and the innovative parent system. Moreover, their progeny is challenging to select excellent offspring varieties.

5.3.1 Directly Crossing of Hybrids and *S. spontaneum* and Its Effect of Plant Improvement

S. spontaneum is the wild *Saccharum* species, making significant contributions to sugarcane breeding and producing many excellent varieties. But with different hybridization types, the breeding effect differed. The hybridization utilization between hybrids and *S. spontaneum* is relatively early. As many as 29 combinations of hybridization had been reported; only 4 combinations successfully bred varieties (Table 5-2). However, the varieties generated by this crossing way have not impacted sugarcane production and parent utilization. The breeding effects are not as sound as that of the symmetric hybridization of the original F_1 generation. There are two main problems. The first one is poor hybridization symmetry. The second is severe inbreeding and networking of existing hybrids. Table 5-2 shows several successful hybridization combinations based on crossing hybrids with *S. spontane*um. There are too many hybridization combinations used to improve parents, while few varieties have been bred. The reason may be mainly due to the worse symmetry and the severe networking.

Table 5-2 Method and effect of direct hybridization between hybrid× *S. spontaneum*

Clones	Hybrid combination	Location	Offspring varieties
Mol1928	B6308×*S. spon*/ Co	Mauritius	H32-6705, H41-3340
B37256	Ba11569×*S. spont*/Co	Barbados	B45181, CP43-64, CP63-588
CYR80-189	POJ3016×*S. spont*/Manhao	Ruili, China	CYZ99-155
PT70-4255	F152×66S82	Pingdong, Taiwan, China	ROC16, ROC23, CGT02-901, CGT03-1403

Source: Wu et al., 2014.

5.3.1.1 Utilization of asymmetric hybridization

(1) Improvement of Mol1928 and its offspring

The germplasm Mol1928 was selected by crossing the Indian *S. spontaneum* with hybrid B6308 in Mauritius. Mol1928 was then back-crossed twice with UD50 and H32-8560 to produce the offspring H32-6705 and H41-3340, respectively. There was no genetic crossover in its pedigree (the consanguinity of B6308 is unknown), but the way of hybridization was extremely asymmetric (Figure 5-16). The hybridization between the original *Saccharum* species and the hybrids occurred many times, including the hybridization between Kansar (*S. barberi*) and POJ213, Uba (*S. sinense*) D1135, etc. Moreover, H32-8560 was the F_1 generation of 'Sugarcane King' POJ2878, but its offspring showed no outstanding production and parental utilization performance. As shown in Figure 5-16, the main reason lies in the asymmetry of the hybridization mode. So the production area and application of the bred variety H41-3340 is not outstanding. There is no report of new varieties bred by using H41-3340 as parents.

(2) Improvement of B37256 and its offspring

The germplasm B37256 was selected by crossing the Indian *S. spontaneum* with the hybrid Ba11156 in Barbados. CP1165, which bred many excellent varieties, was used to improve the traits of B37256. Although there is no crossover between the parents of CP57-120, the traits of its offspring were not outstanding. Meanwhile, CL54-190, the third generation of POJ2878, was used for breeding variety CP63-588, which has poor production performance. It may be due to an unclear genetic relationship and asymmetric hybridization mode of its paternal CP57-120 (Figure 5-17). No variety has been bred by using B37256 as the parent.

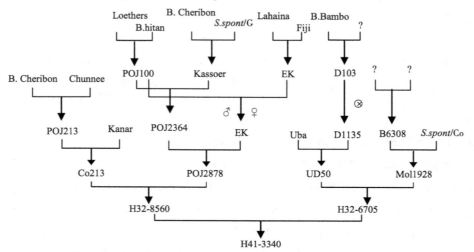

Figure 5-16 The pedigree chart and improvement of Mol1928

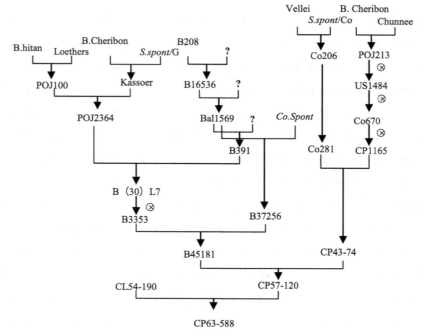

Figure 5-17 The pedigree chart and improvement of CP63-588

5.3.1.2 Method of sugarcane improvement with serious parental networking

(1) Improvement of CYR80-189 and its offspring

In Ruili National Inland Sugarcane Hybrid Breeding Station, China (RSBS), the basic germplasm CYR80-189 was selected by crossing Yunnan *S. spontaneum* with POJ3016. Although the new consanguinity of *S. spontaneu*m was introduced into the offspring, the backcross parents, such as POJ3016, F134, F172, and ROC10, were all the offspring of POJ2878 (Figure 5-18), representing repeated usage of POJ2878 for hybridization and back-crossing in different generations. The networking of parents becomes more serious. The current cultivar CYR99-155 has high sugar content in Yunnan Baoshan Shangjiang Sugar Mills, but its adaptability is imperfect and susceptible to slight rot.

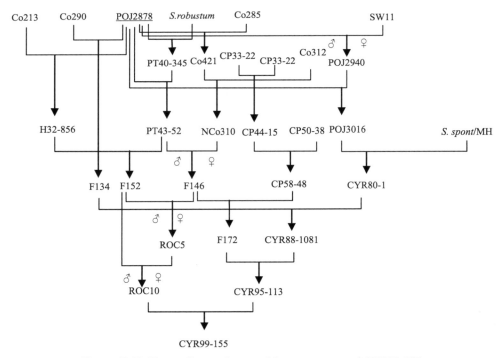

Figure 5-18 The pedigree chart and improvement of CYR99-155

(2) Improvement of PT70-4255 and its offspring

In Taiwan Province of China, the innovative germplasm PT70-4255 and PT74-575 were bred by crossing *S. spontaneum* (66S82) with F152 variety (Figure 5-19). Two varieties were finally screened out. However, both female parents, F171 and F166, are the offspring of POJ2878, repeatedly used in different generations. The symmetry of hybridization is also wrong.

In Sugarcane Research Institute, Guangxi Academy of Agricultural Sciences (GXSRI), ROC23 was hybridized with CP84-1198 to produce CGT02-901 and CGT03-1403, which genetic relationship also showed some crossover. The offspring of the improved parents were bred into some better varieties, such as ROC16 and ROC23. These varieties were also benefited from the consanguinity of PT43-52. PT43-52, the offspring of *S. robustum*, was used for breeding F145 and F152 directly. Due to the limitation of consanguinity, the traits and parental utilization of these varieties showed limited potential.

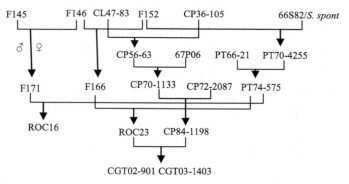

Figure 5-19 The pedigree chart and improvement of GT02-901 and GT03-1403

5.3.2 Sugarcane improvement by Directly Crossing Hybrids with *S. robustum*

S. robustum is mainly manifested with thick stem diameter and strong lodging resistance, making significant contributions to sugarcane breeding and producing many excellent varieties. The hybridization utilization between hybrid and *S. robustum* is very early. Since the 1930s, this hybridization type was carried out in the province of Taiwan, yunnan, and Hainan of China, as well as Hawaii of USA. A total of 8 hybridization combinations (Table 5-3) have been collected in this section, but only 3 combinations have been reported to breed varieties successfully, 2 of which are made in province of China, and 1 in Hawaii, USA (Table 5-3, Figure 5-20, Figure 5-21). There were more *S. robustum* species used to improve the traits of parents with worse symmetric crossing and more serious networking, but no varieties were bred.

Table 5-3 Improvement and hybrid utilization of hybrids and *S. robustum*

Clones	Hybrid combination	Location	Offspring varieties
PT40-345	POJ2875×*S. robustum*	Pingdong, Taiwan, China	Many
PT40-388	F108×*S.robustum*	Pingdong, Taiwan, China	Many
H32-5774	Mol1231×*S. robustum*	Hawaii, USA	Many
PT39-461	POJ2883×*S. robustum*	Pingdong, Taiwan, China	None
PT40-203	POJ2725×*S. robustum*	Pingdong, Taiwan, China	None
CYC 96-49	IJ76-315×NG77-004	Yacheng, Hainan, China	None
CYC80-142, CYC80-143	S17×*S. robustum*	Yacheng, Hainan, China	None
CYR03-403	CMT70-611×57NG208	Ruili, Yunnan, China	None

Source: Wu et al., 2014.

5.3.2.1 Crossing improvement with distinct consanguinity and good symmetry

(1) Improvement of PT40-345 and its offspring

Hybridization of POJ2875 and *S. robustum* was used for breeding PT40-345 germplasm, and it was then backcrossed with POJ2878 to obtain germplasm PT43-52 (Figure 5-20) in Taiwan Province of China. POJ2875 and POJ2878 are complete sibling lines, and both are excellent parents. This method is the most successful hybridization utilization of *S. robustu*m so far. It has been reported that their offspring had been bred in Province of China and many countries worldwide. There are more than 50 varieties cultivated by PT43-52, including F140−F178 and 27 varieties developed in Taiwan Province of China. 4 varieties developed in

Taiwan Province of China, and other varieties have many promotions and essential applications in the mainland of China. In addition, more than 20 offspring of PT43-52 have been bred in Reunion, Philippines, Malaysia, Australia, and China. Many offspring have been bred by using F146, F152, and F141 as parents among these varieties.

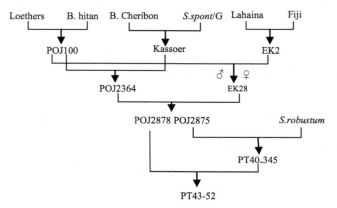

Figure 5-20 The pedigree chart and improvement of PT43-52

5.3.2.2 Crossing improvement with serious asymmetry

Improvement of H32-5774 and its offspring: The basic germplasm H32-5774 was screen out by hybridizing Mol1231 with *S. robustum* in Hawaii Sugarcane Breeding Station. It was then crossed with H27-8101, F_1 of Badila, to generate the H34-1874 variety. H34-1874 was further backcrossed with H32-856 to breed H37-1933 (Figure 5-21). The germplasm Mol1231 is of unknown origin. Multiple rounds of asymmetric hybridization generated H27-8201. H109 is the F_1 hybrid of two *S. officinarum* (*Lahaina* and US1494). 25C14 and H27-8101 are the F_1 generations of Y.Cheribon (*S. officinarum*) and Badila (*S. officinarum*), respectively. There is no change of generations in the three offspring, resulting in extremely poor asymmetry of consanguinity. The continuous hybridization utilization of H37-1933 bred H49-5, H49-4360, H59-3775 in Hawaii, and PSR98-38 in The Philippines. Due to the consanguinity relationship, there were no reports of widespread promotion of these varieties in the local area. Moreover, they didn't breed any new variety.

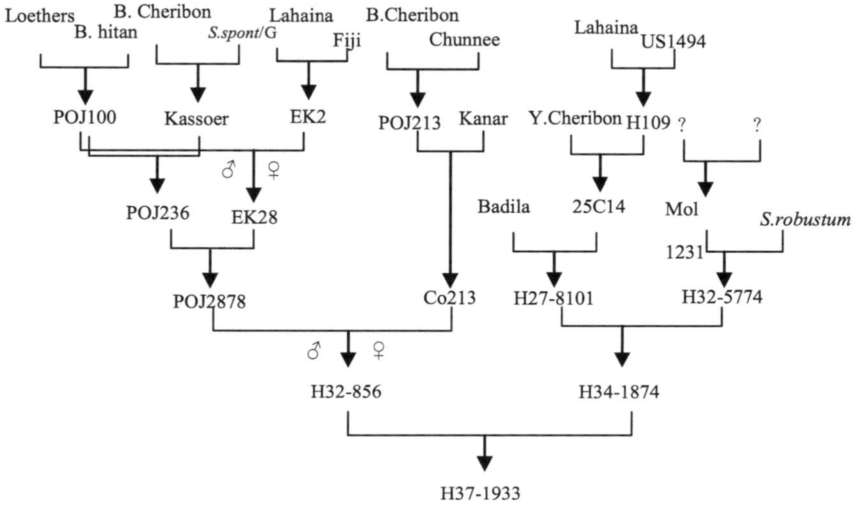

Figure 5-21 The pedigree chart and improvement of H32-5774

5.3.2.3 Parents improvement with serious networking and their breeding effect

Improvement of PT40-388 and its offspring: PT40-388 was bred by crossing F108 with *S. robustum* in Taiwan Province of China. It was used to cross with Co419 to generate CYC73-226. F108 and Co419 are both worldwide varieties. However, the traits of PT40-388 were not good as PT43-52 on account of more severe networking of parental consanguinity. The continuous hybridization of PT40-388 bred CMT96-1409, CMT99-596, and CYT79-177 in the mainland of China. The number and the extension areas of its offspring varieties are limited.

5.3.3 Direct Hybridization and Utilization of Hybrids with *S. officinarum*

Hybridization between hybrids and *S. officinarum* began very early. Since the setting of sexual hybridization, this method had been carried out in many countries. In this paper, Wu caiwen summarized 17 times this mating system, 7 of which bred varieties effectively (Table 5-4). There are two different ways of hybridization, including *S. officinarum* × hybrids and hybrids × *S. officinarum*. Their hybrid offspring of both combinations did not significantly influence sugar production or sugarcane breeding. There are two main problems in these hybridization types. One is the poor symmetry of crossing. The other is the networking of parental consanguinity. Table 5-4 represents several improved parents and varieties successfully bred by crossing hybrids with *S. officinarum*. The hybridization combinations which bred no variety also have a severe disadvantage in symmetry and parental networking.

Table 5-4 Improvement and hybrid utilization of hybrids *and S. officinarum*

Clones	Hybrid combination	Hybrid type	Location	Offspring varieties
25C14, 26C270	Y.Cheribon×H109	*S. officinarum*×F_1	Cuba	Many, H31-1389、H49-3646, etc.
Co453	B.Cheribon×Co285	*S. officinarum*×F_1	India	Many, Co785, CR67274, My5715, etc.
POJ2221, POJ2222, POJ2320, POJ2191	B.Cheribon×Kassoer	*S. officinarum*×F_1	Java, Indonesia	Fewer, POJ2628, Mex73-206
Co214	S.Mauritius× M4600	*S. officinarum*×F_1	India	Many, Co320, Co327, R331, CoTo,.etc.
CYC71-370	Vietnam Niucane×-CYC58-47	*S. officinarum*×F_1	Yacheng, Hainan, China	CGN81-1035
CYC65-621	Badila×Kassoer	*S. officinarum*×F_1	Yacheng, Hainan, China	None
CYC88-196	Waie×CYR80-189	*S. officinarum*×F_1	Ruili, Yunnan, China	None
CYC74-228	WanNing Lazhe×-CYC62-40	*S. officinarum*× F_2	Yacheng, Hainan, China	None
Comus	Oramboo×Q813	*S. officinarum*×hybrids（?）	Australia	None
MQ28-13	Badila×Q813	*S. officinarum*×hybrids（?）	Mackay, Australia	None

				continued
Clones	Hybrid combination	Hybrid type	Location	Offspring varieties
Co223	Chittan×M1515	S. officinarum×hybrids（?）	India	None
CYC97-27	Korp×CYC96-68	S. officinarum×F_1*	Yacheng, Hainan, China	None
CYC97-38, CYC97-40	50Uahiapele×-CYC95-41	S. officinarum×F_1*	Yacheng, Hainan, China	None
POJ2747	POJ2628×Lahaina	S. officinarum×F_2	Java, Indonesia	POJ2927, POJ2928, POJ2929
CYC59-818	S17×Badila	S. officinarum×F_3	Yacheng, Hainan, China	CYZ71-388, CLH78-337
Damon	Trojan×Oramboo	S. officinarum×F_2	Australia	None
PT40-196	POJ2725×Badila	S. officinarum×F_3	Pingdong, Taiwan, China	None

Note: * represent the F_1 hybrid of E. arundinaceum

Source: Wu et al., 2014.

5.3.3.1 Methods and effects of hybrid utilization between S. officinarum and the progeny of Saccharum interspecific hybridization

(1) The way and effect of hybridization utilization between S. officinarum and F_1 hybrids of Saccharum interspecific hybridization

In the hybridization of S. officinarum and hybrids, the genetic relationship is generally clear but asymmetric when the hybrids are the F_1 of the original Saccharum species. The F_1 hybrids, no matter origin from the intraspecific crossing of S. officinarum or interspecific hybridization of S. officinarum and S. spontaneum, still belong to the F_1 generation of S. officinarum (Figure 5-22). Such a combination has a high probability of breeding sugarcane varieties. Wu et al. collected 7 hybrid combinations, 5 of which bred varieties successfully. For example, 25C14, 26C270, H27-8101, Co453, POJ2191, POJ2221, POJ2222, POJ2320, Co214, CYC71-370, and CYC79-257 were successfully bred by this hybridization mode. Among these varieties, H27-8101 was quite extraordinary. The offspring of *S. officinarum* was backcrossed with different *S. officinarum* species two times. The generation of the offspring didn't change, but different kinds of *S. officinarum* clones and the proportion of consanguinity varied. Due to the different *S. officinarum* species, there is no crossover of consanguinity, which is the most significant advantage of these varieties.

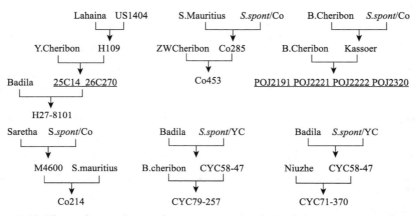

Figure 5-22 The pedigree charts of some returned F_1 hybrids by crossing *S. officinarum* with original F_1 hybrids

On the other hand, CYC65-621 and CYR88-196 were created by the intraspecific hybridization of *S. officinarum* species, but they didn't breed any new varieties. It may seem due to insufficient rounds of hybridization. However, it is likely to cause asymmetric consanguinity and common traits based on the successful hybridization utilization cases. It is not necessary to spend a lot of time to continue this way of hybridization.

(2) The way and effect of hybridization utilization between *S. officinarum* and F_2 hybrids (or hybrids with unknown consanguinity)

The only hybridization combination between *S. officinarum* and F_2 generation produced CYC74-228 was collected (Table 5-4). However, it did not successfully develop offspring varieties. As shown in Figure 5-23, the consanguinity proportion of CYC74-228 is seriously asymmetric and complicated. Although its parents include world-famous parents, including Co290, POJ2878, and F134, CYC74-228 cannot breed any varieties by continuous hybridization. For Comus, MQ28-13, Co223, and other clones with unknown consanguinity, it is not convenient to make comments. Given its early time usage and none excellent offspring, the reason is nothing more than two. One is the poor symmetric way of crossing. The other is the severe networking of genetics. Their pedigree charts should be similar to CYC74-228.

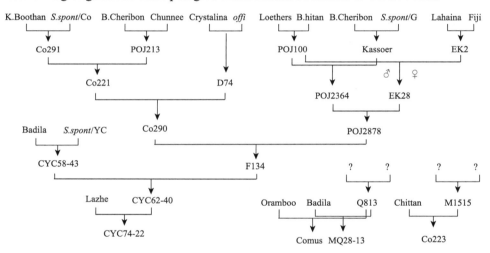

Figure 5-23 The pedigree charts of CYC74-228, Comus, MQ28-14 and Co223

5.3.3.2 Methods and effects of hybridization utilization between *S. officinarum* and *Saccharum* relative genus

The hybridization combinations between *S. officinarum* and *Saccharum* relative genus collected in this section were mainly performed by crossing *S. officinarum* and *E. arundinaceus*. Their offspring were continued to cross back with *S. officinarum*, resulting in the same F_1 generation of *S. officinarum*. Table 5-4 showed three innovative materials, namely CYC97-27, CYC97-38, and CYC97-40. They were derived from two combinations. Their pedigree charts were shown in Figure 5-24, presenting clear consanguinity. According to our experience, the hybrid offspring can further breed varieties if the *E. arundinaceus* is changed to *Saccharum* wild species (*S. spontaneum* or *S. robustum*). But the *E.arundinaceus* is the relative genus of *Saccharum* species; there was no report that any variety was bred by using *E. arundinaceus* as a parent. Whether the hybrids of *S. officinarum* and *E. arundinaceus* could further produce excellent sugarcane varieties still needs to be confirmed through more time.

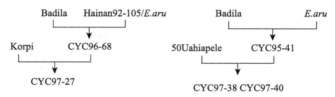

Figure 5-24 The pedigree charts of the hybrid offspring of *E.arundinaceus*

5.3.3.3 Methods and effects of hybridization utilization between *S. officinarum* and *Saccharum* relative genus

The hybridization combinations between *S. officinarum* and *Saccharum* relative genus collected in this section were mainly performed by crossing *S. officinarum* and *E. arundinaceus*. Their offspring were continued to cross back with *S. officinarum*, resulting in the same F_1 generation of *S. officinarum*. Table 5-4 showed three innovative materials, namely CYC97-27, CYC97-38, and CYC97-40. They were derived from two combinations. Their pedigree charts were shown in Figure 5-24, presenting clear consanguinity. According to our experience, the hybrid offspring can further breed varieties if the *E. arundinaceus* is changed to *Saccharum* wild species (*S. spontaneum* or *S. robustum*). But the *E.arundinaceus* is the relative genus of *Saccharum* species; there was no report that any variety was bred by using *E. arundinaceus* as a parent. Whether the hybrids of *S. officinarum* and *E. arundinaceus* could further produce excellent sugarcane varieties still needs to be confirmed through more time.

5.3.3.4 Genetic relationship of the varieties which fail to breed new varieties

Among the hybrids produced by the hybrids × *S. officinarum*, Damon and PT40-196 were fail to breed new varieties. The dendrogram shows that Damon and PT40-196 belong to the F_3 and F_4 generations of *S. officinarum*, respectively (Figure 5-26). Compared with the above hybrids with newly bred varieties, the reasons should be due to the advanced generations and the complex consanguinity. By analyzing this hybridization mode, the obtained germplasm should also be able to breed new varieties. But the traits of the offspring would not be outstanding. So this way of hybridization should be avoided as far as possible.

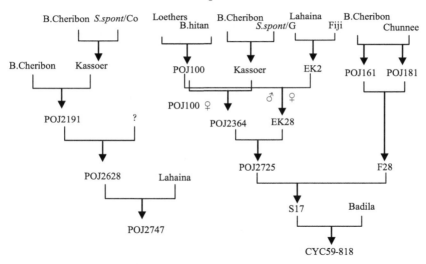

Figure 5-25 The pedigree charts of POJ2747, CYC59-818

Table 5-5 Improvement and hybrid utilization of hybrids and *S. sinense*

Clones	combination	Type	Generation	Location	Offspring varieties
UD50	Uba×D1135	*S. sinense*×Hybrids	F_1	Hawaii, USA	Many, such as Mo1928, H33-6705, H78-0698.
H24-4399	Uba×H450	*S. sinense*×Hybrids	F_3	Hawaii, USA	H32-1063
H28-4772	Uba×H456	*S. sinense*×Hybrids	F_3	Hawaii, USA	Many, such as H31-2346, H70-6957

continued

Clones	combination	Type	Generation	Location	Offspring varieties
CP726	POJ213×Uba	S. sinense×Hybrids	F_1	Canal point, USA	Many, such as CL41-142, CP71-1038, Hocp92-624, ROC22, CYN81-762
CYC57-36	zhucane×F134+Co331	S. sinense×Hybrids	F_2	Yacheng, Hainan, China	None
CYC73-92	zhucane×CYC58-63	S. sinense×Hybrids	F_4	Yacheng, Hainan, China	None

Source: Wu et al., 2014.

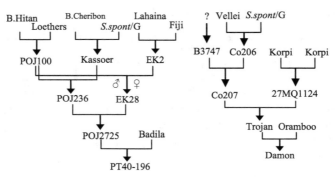

Figure 5-26 The pedigree charts of PT40-196 and Damon

The results showed that: Firstly, in the hybridization between *S. officinarum* and hybrids, the hybrids derived from *S. officinarum* and *Saccharum* species were easier to breed new varieties than that origin from *S. officinarum* and *Saccharum* relative species. The lower the generation of the hybrids has the more possibility to breed new varieties. The F_1 hybrids could be continued hybridization to breed varieties more successfully than F_2 or higher generations. The varieties obtained by crossing *S. officinarum* with *Saccharum* relative species have not been reported. Secondly, in the hybridization of hybrids and *S. officinarum*, the clearer parental consanguinity and the less genetic crossover, the offspring can more easily breed new varieties. Both the hybridization of *S. officinarum* × hybrids and hybrids × *S. officinarum* can cultivate sugarcane varieties. However, the bred varieties have limited production performance and promotion areas. It is recommended not to spend many workforces, material resources, and time researching this type of hybridization.

5.3.4 Methods and Effects of Sugarcane Improvement by Direct Hybridization between Hybrids and *S. Sinense*

The direct hybridization between hybrids and *S. sinense* is early, and the breeding effect is significant. This type of hybridization has been conducted in both China and the United States. There are a total of 6 hybrid combinations collected by Wu Caiwen, 4 of which Uba (*S. sinense*) was used (Table 5-5).

5.3.4.1 Methods and effects of hybrid utilization between Uba and hybrids

In the reported hybridization utilization between hybrids and Uba, all hybrids are the offspring of interspecific hybrids. There are two ways of hybridization between Uba and hybrids, including Uba× hybrids and hybrids × Uba. A large number of excellent varieties have been bred in both ways. Therefore, it is speculated that Uba, an *S. sinense* species, should not be the distant interspecific offspring of *Saccharum* relative wild species and *S. officinarum*, but the natural interspecific offspring of *S. officinarum* and *Saccharum* wild species.

(1) Methods and effects of hybrid utilization with Uba×hybrids

The earliest and most successful utilization of Uba is the Hawaii Sugarcane Breeding Station, United States. Three progeny clones, UD50, H24-4399, and H28-4772, were bred by crossing Uba with D1135, H450, and H456, respectively (Table 5-5). The original *Saccharum* species of D1135, H450, and H456, were analyzed, including F_1 and F_3 (Figure 5-27). According to Table 5-5, UD50 is the F_1 generation of Uba, and the offspring breed many varieties. Irvine (1999) thinks that Uba is the hybrid offspring of *S. officinarum*×*S. spontaneum*. Therefore, the hybrid of Uba and D1135 belongs to peer crossing, so the bred varieties are many, and the characters are excellent. The parents of H24-4399 and H28-4772 are identical in blood relationships. Their female parent is Uba, and their male parent is a sister line from the same combination. They are all F_3 generation (D1135 is the selfing progeny of D103, so D1135 is also the F_1 generation of R.bamboo). Due to the asymmetry of blood relationships, fewer varieties are bred.

(2) The essential hybrid combination of hybrid × Uba

The only combination POJ213×Uba, collected by Wu Caiwen, was made in the Hawaii Sugarcane Breeding Station of the United States. The offspring CP726 was successfully cultivated. With the repeated hybridization and backcross of CP726, many varieties were bred in the United States, Brazil, Thailand, and China (Table 5-5). Some varieties showed good production performance and were widely planted in China's mainland. Why the essential hybrid combination of hybrids × Uba can breed so many excellent varieties? Wu Caiwen convinced the previous research, which supposed Uba as the progeny of *S. officinarum*× *S. spontaneum*. As shown in Fig 5-28, the hybridization of POJ213 and Uba is symmetric, which is easy to breed excellent varieties. Otherwise, if Uba is the hybrid offspring of *S. officinarum* × *M. sinensis*, its flowering competence, and pollen sterility will not be so lovely. It can't be able to breed such many excellent offspring as the male parent.

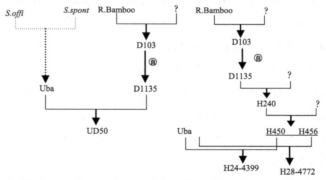

Figure 5-27 The pedigree charts of the offspring of *S.sinense* and hybrids

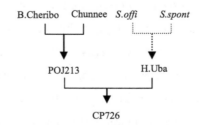

Figure 5-28 The pedigree charts of CP726

5.3.4.2 Methods and effects of crossing hybrids with zhucane and Niucane

The hybridization and utilization of zhucane (*S. sinense*) and Niucane (*S. sinense*) began in 1957. It was not until 1973 that Tanzhou zhucane and Guangxi zhucane were hybridized with 4 hybrids in Hainan Sugarcane Breeding Station. Two basic hybridizations were completed, and the hybrid offspring CYC57-36 and CYC73-92 were obtained (Table 5-5). In the two combinations, Both zhucane and Niucane were used as the female parent, mainly because of insufficient pollen sterility. Their male parents were F_2 and F_4 generations, respectively. During multiple years and different types of hybridization utilization as male parents, zhucane and Niucane didn't breed any new varieties. It is not only because of asymmetric crossing (Figure 5-29) and related to the origin of the two *S. sinense*. Grassl believed that *S. sinense* was the natural hybrids of *S. officinarum* and *MiscanthuS. sacchariflorus*. The distant hybrid offspring exhibited insufficient pollen sterility and can only be used as the female parent. Wu Caiwen agreed with Grassl's opinion that zhucane and Niucane were the progeny of *S. officinarum* and *M. sinensis*. There are no reports on breeding varieties by distant hybridization utilization of *Saccharum* relative species in the world.

In short, *S. sinense* played a vital role in the historical process of sugarcane breeding. Due to the successful importation of the consanguinity of *S. sinense* into modern sugarcane varieties, significant progress has been made in the stress resistance and adaptability of modern sugarcane varieties. The advantages of *S. sinense* have promoted the development of the modern sugarcane industry, such as resistance to drought, barrenness and cold, ratooning solid, good tillering, and developed root system.

Firstly, the breeding effect showed that Uba could be used as both male and female parents. It should be the interspecific hybrid offspring of *S. officinarum* and *S. spontaneum*, symmetric hybridization between F_1 generations.

Secondly, Other kinds of *S. sinense*, such as zhucane and Niucane, are not accessible to blossom and poor pollen infertility. It is because that they are the hybrid offspring of *S. officinarum* and *M. sinensis*. Their effects on hybridization utilization are far less than Uba.

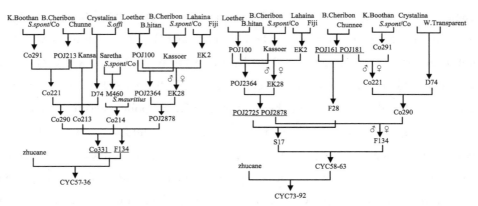

Figure 5-29　The pedigree charts of CYC57-36 and CYC73-92

5.3.5　Methods and Effects of Sugarcane Improvement by Direct Hybridization between Hybrids and *S. Barberi*

S. barberi is suitable to plant in subtropical and temperate regions. It is mainly characterized by prematurity, high sugar content, high fiber content, drought tolerance, barrenness tolerance, extensive cultivation, developed root system, more tiller, good ratooning, and muscular stress resistance. The chromosome numbers of *S. barberi* are $2n = 82-124$, the same as that of *S. sinense*. Irvine (1999) believed that *S. barberi* was a natural hybrid offspring of *S. officinarum* × *S. spontaneum*. But Grassl thought that *S. barberi* was the progeny of *S. officinarum* ×*M. sinensis*. Currently, there are two hybridization methods, namely hybrids × *S. barberi* and *S. barberi* × hybrids. There are two popular *S. barberi* used, namely Chunnee and Kansar, used as male and female. Many excellent offspring varieties have been bred. Based on the traits of offspring, It was confirmed to Irvine's research that *Chunnee* and Kansar were the progeny of *S. officinarum*×*S. spontaneum*. Due to the difficulty in flowering, pollen dysplasia, and low fertility of other *S. barberi*, the author Wu Caiwen speculated that these *S. barberi* were not the natural hybrids of *S.*

officinarum × *S. spontaneum*, but the natural hybrids of *S. officinarum* × *Saccharum* related species. They have little contribution to sugarcane breeding.

5.3.5.1 The essential hybridization of *S. barberi* × hybrids

Author Wu Caiwen collected only one effective hybrid combination of S. barberi × hybrid (Table 5-6). In Java Sugarcane Breeding Station, Indonesia, Chunnee was used to cross with the POJ100 (F_1 hybrid of original *Saccharum* species), and the offspring were successfully selected out POJ385 (Figure 5-30). Many varieties were successfully bred using POJ385 for further hybridization or back-crossing in India, Taiwan province of China, the USA, and South Africa. However, no breakthrough variety was produced in the offspring of Chunnee, which might be related to genetic crossover in the parental consanguinity *of S. officina*rum.

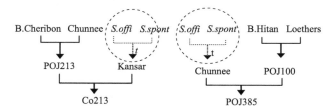

Figure 5-30 The pedigree charts of Co213 and POJ385

5.3.5.2 The essential hybrid combination of hybrids × Kansar

Wu Caiwen collected only one effective hybrid combination of hybrid × *S. barberi* (Table 5-6). In Coimbatore SugarcaneBreeding Station, India, POJ213, a hybrid of the F_1 generation of *Saccharum* species, was hybridized with Kansar (*S. barberi*), and then the offspring Co213 was screened (Figure 5-30). According to Irvine's opinion, Kansar should be a natural hybrid of *S. officinarum*×*S. spontaneum*. So the bi-parents of Co213 were the equivalent generations of original *Saccharum* species. Co213 was considered a worldwide parent. By continued hybridization of Co213, many worldwide varieties were developed, such as Co290, F134, NCo310, and Co419. According to Irvine's study, Kansar was also a natural hybrid of *S. officinarum*×*S. spontaneum*.

Table 5-6 Improvement and hybrid utilization of hybrids and *S. barberi*

Clones	Combination	Type	Generation	Location	Offspring varieties
POJ385	Chunnee×POJ100	*S. barberi* × hybrid	F_1	Java, Indonesia	Many, such as Co301, F105, CL41-114, Co1148, Co678, N10, N55-805 etc..
Co213	POJ213×Kansar	hybrid×*S. barberi*	F_1	Coimbatore, India	Many, such as Co290, Co419, Co312, H32-856, H37-1933, H41-3340, NCo310, F134, F146, F172, F160, ROC5, ROC10, K82-65, K82-83 etc.

Source: Wu et al., 2014.

5.3.6 Sugarcane Parent Improvement and Its Breeding Effects

There are many hybrid types used to improve the parents' traits. But the number of varieties that improved parents bred is not large. The quality, production performance, utilization, and popularization areas of these bred varieties are limited, making little contribution to sugarcane breeding and the cane sugar industry. The common characteristics of the improved parents were as follows:

Firstly, the breeding efficiency is not high. Among the direct hybridization utilization with *S. spontaneum*, only 4 of 29 hybrid combinations successfully bred varieties.

Secondly, the offspring of the improved parents didn't exhibit obvious breakthroughs in traits and production, and continued hybridization of the offspring produced little new variety.

Thirdly, in the sugarcane parents' improvement, there is a particular case of basic hybridization. Hybrids were used to cross with *S. robustum*, and the offspring were continued to backcross with hybrids. PT43-52, which was bred through this strategy, showed a better effect on sugarcane breeding than the hybridization of *S. officinarum* with *S. robustum*. Further, it is needed to confirm whether crossing *S. officinarum* with the offspring of *S. robustum* is the optimal utilizing way of *S. robustum*.

Forthly, the Kansar, Chunnee, and Uba are hybrids of *S. officinarum* and *S. spontaneum*. Using this germplasm was slightly different from that of *S. officinarum* in breeding breakthrough sugarcane varieties. The direct hybridizations between these materials with the F_1 generation of the original species are peer-based hybridization. Other kinds of *S. barberi* and *S. sinense* with disadvantages in flowering, abnormal pollen, and low fertility should be the offspring of *S. officinarum* and *M.sinensis*.

5.4 Influence of Hybridization Mode on the Traits of Offspring

By analyzing the different essential hybridization and the effects on the traits of the offspring, it was found that the hybridization mode had a significant impact on the offspring's characters. It indicates that selecting the correct hybridization mode is no less important than selecting the right parents. Failure to select the optimal materials and hybridization methods is equivalent to the waste of workforce, material resources, and financial resources. It is challenging to cultivate excellent sugarcane varieties, the least of breakthrough varieties.

5.4.1 The Independent Parent System and The Traits Performance of The Progeny

Over the past 100 years in sugarcane hybrid breeding, there have been many intraspecific hybridization methods within original *Saccharum* species. But there are three effective types of hybridization, including *S. officinarum* × *S. officinarum*, *S. officinarum* × *S. barberi*, and *S. officinarum* × *S. spontaneum*. In detail, there were 11 hybridization combinations of *S. officinarum* × *S. officinarum*, 7 of which were taken in the early sugarcane breeding program. The time of crossing *S. officinarum* with *S. barberi* was earlier, and one of the three collected combinations produced excellent offspring varieties. Twenty hybrid combinations of *S. officinarum* × *S. spontaneum* were gathered, 14 of which were taken in the early sugarcane breeding program. 10 combinations bred varieties successfully at a ratio of 71.4%. All the breakthrough cultivar was bred utilizing peer-based hybridization. According to the report, sugarcane breeders have recently carried out 10 hybrid combinations similar to the *Saccharum* species. As long as we pay attention to the development of peer-to-peer hybridization without blood cross and lay the foundation for the cultivation of a new independent parent system, the future generations are expected to have breakthrough varieties; However, if the cross mode is asymmetric and the heterogeneity of pairing mode is not significant, it is difficult to cultivate breakthrough sugarcane varieties.

5.4.2 Breeding Parents and Their Performance of Hybrids Between F_1 and Higher Generations of The *Saccharum* Species

Some undesirable traits of the existing parents can be improved through the introduction of new genetic resources. The more new genetic resources are input, the better the improvement of parental characters will be, and the greater the possibility of breeding breakthrough varieties should be. A new independent parent system is developed when enough consanguinity of new parents is introduced by symmetric hybridization. If the F_1 generation of the original *Saccharum* species is hybridized asymmetrically, the offspring will contain much consanguinity of two new species. Then there will be more improvement in the traits of the varieties. If the genetic bases of parents are distant, which usually show significant differences in traits and a high degree of heterosis, the offspring will have better production performance. It was found that the hybridization of F_2 generation and BC_1 generation was more likely to breed excellent varieties than that of F_3 generation, BC_2 generation, or higher generation. Previous research tells us that sugarcane breeders should adopt the new original *Saccharum* species to broaden the genetic base of modem cultivar. Meanwhile, it should not be easy to carry out the asymmetric way of hybridization. Because these hybridization modes are ineffective in sugarcane breeding, the offspring are less likely to break through their traits.

5.4.3 Breeding Innovation Parents and Their Performance by Direct Hybridization of Original *Saccharum* Species and Higher Generation Hybrids

The analysis shows the direct hybridization of original *Saccharum* species and higher generation hybrids will only produce improved sugarcane varieties with poor economic traits. No matter what the original *Saccharum* species are and how excellent the quality of the hybrids is, the success rate of cultivating the excellent sugarcane varieties is low. Moreover, the characters of the improved sugarcane varieties will not be outstanding. The improved sugarcane varieties have a smaller promotion area and contribute to sugarcane production, which is significantly inferior to the innovative varieties bred by the original F_1 hybrid. It is also inferior to the worldwide parents and the independent parent system in sugarcane breeding. Therefore, this method is one that sugarcane breeders should avoid as far as possible.

In conclusion, the progeny of independent parents, innovative parents, and improved parents differ significantly in many characteristics. But the original *Saccharum* species are adopted almost the same. The major difference is the symmetry of basic hybridization. Whether it is an independent parent system, innovative parent, or improved parent, they all breed new offspring varieties. The dendrogram of offspring is an inverted triangle. The better the symmetry is, the better the traits of offspring will be. The clearer the genetic relationship is, the more significant the breakthrough of varieties will be. The parents created by innovative and improved parents system didn't breed worldwide varieties or new worldwide parents. Their offspring showed minor breakthroughs of traits, and the promotion areas of these varieties were also limited. Therefore, sugarcane breeders should select the parents of crossbreeding and analyze the genetic relationship of parents to improve the efficiency of sugarcane breeding.

5.5 Advances in Breeding New Independent Parents of Sugarcane by the Peer to Peer Hybridization in Yunnan

5.5.1 The Idea of Peer-to-peer Hybridization to Breed Independent Parent System

The parents are the material foundation stone for breeding breakthrough sugarcane varieties (Chen et al., 2011; Wu, 2005; Wu et al., 2013; Wu, 2018). Over the past ten years, sugarcane crossbreeding workers worldwide have done much work in germplasm innovation and crossbreeding utilization of parents. Thousands of hybrids have been selected, and a large number of parents have also been created. However, because the hybridization has not been separated from the two-parent systems of "POJ" and "Co", As a result, there was no significant breakthrough in the traits of sugarcane varieties developed later (Chen et al., 2011; Wu, 2005; Wu et al., 2013; Wu, 2018).

Why are there many sugarcane germplasm resources, and the problems of sugarcane parents' narrow genetic basis, Networking of kinship, and inbreeding of sugarcane parents, have not been solved for a long time? Since the 1990s, domestic sugarcane breeding organizations have carried out extensive scientific and technological cooperation with major global sugarcane scientific research institutions. Some scientific and technological personnel have successively gone to major sugarcane-producing countries and institutions in the world for investigation, study, and cooperative research and have mastered much first-hand information on sugarcane crossbreeding (Chen et al., 2011; Wu, 2005; Wu et al., 2013; Wu, 2018).

Based on absorbing the experience and lessons of sugarcane hybrid breeding at home and abroad, adjusting the breeding ideas, and innovating the breeding methods, the Sugarcane Research Institute of Yunnan Academy of Agricultural Sciences (YSRI) for the first time proposed the idea of using the independent parent system to cultivate breakthrough sugarcane varieties through peer to peer hybridization (Wu, 2005; Wu et al., 2013; Wu, 2018).

5.5.2 Utilization of New Ancestral Species in Basic Hybridization

Yunnan is the second-largest sugar cane production base in China and one of the regions with the richest wild sugarcane resources in the world (Fan et al., 2001; Jing et al., 2008; Bian et al., 2015; Liu et al., 2014; Tao et al., 2011). To promote the collection and innovative utilization of sugarcane germplasm resources, China has successfully established the National Germplasm Repository of Sugarcane and the Ruili Inland Hybrid Breeding Base in Yunnan 1980s. By the end of the 13th Five-Year Plan of China, relying on the National Germplasm Repository of Sugarcane. YSRI has established the Sugarcane Germplasm Resources Database System, which has the largest number of sugarcane resources. The complete data in China has cataloged 3151 sugarcane resources from 15 species of 6 genera. Phenotypic traits combined with molecular biological technology, a batch of sugarcane genetic materials with excellent characters such as sugar content, yield, disease resistance, and drought resistance were explored and excavated. On the use of germplasm innovation according to the thinking of cultivating new independent parent system, through the resistance evaluation on a large scale, flowering characteristics, and photoperiodic induction research, successively using 21 new species (including *S. officinarum* L., *S. sinense* Roxb., *S.* barberi, Landrace, *S. spontaneum* L., *S. robustum* Brand et. Jewiet) (Table 5-7), to launch a new original basic hybridization between species, Some excellent basic hybrid germplasm of F_1 generation were created, laying a solid foundation for the cultivation of new independent parent system (Jing et al., 2013; Jing et al., 2014; Tao et al., 2014; Tao et al., 1997; Bian et al., 2014; Tian et al., 2017).

Table 5-7 Summary of new independent parents bred by Yunnan peer-to-peer hybridization using new original *Saccharum* species

Sugarcane variety type	Clone of original *Saccharum* species	Quantity
S. officinarum.	48Mouna, 51NG90, Badila, Barwilspt, Zopilata, Canablanca, Ganapathy, 50uahapele, Luohan cane	9
S. sinense	Pansahi, Bailou cane, Guangze bamboo cane, Xuchang chewing cane (Henan, China), Nanjian chewing cane	5
Landrace	Baimei cane, Pupiao chewing cane	2
S. robustum Brand	51NG63, 57NG208	2
S. spontaneum	Yunnan82-59, Yunnan 82-114, Yunnan 82-157, Yunnan 84-268	3

5.5.3 Development and Advantage Analysis of New Independent Parent System by Peer-to-peer Hybridization

By interspecific hybridization of the original *Saccharum* species, we increase the number of generations of peer-to-peer hybridization and pass on more excellent improvements to the sugarcane progeny. A great breakthrough has been made in cultivating new independent parent systems by breaking the narrow consanguinity basis of existing sugarcane varieties, parental consanguinity interleaved use, and the inbreeding of sugarcane parents. A batch F_1 generation excellent innovative germplasm has been created.

Through peer-to-peer hybridization, CYR13-47, CYR14-211, CYR15-55, CYR17-82, and CYR 18-188 were obtained a batch of new F_2 and F_3 parents with a great breakthrough in species (Tao et al., 2020; Yu et al., 2015; Yu et al., 2019a; Yu et al., 2019b; Tao et al.,2011; Tao et al., 2015; An et al., 2008; Hu et al., 2021), and the pedigree of some innovative parents was shown in Figure 5-31.

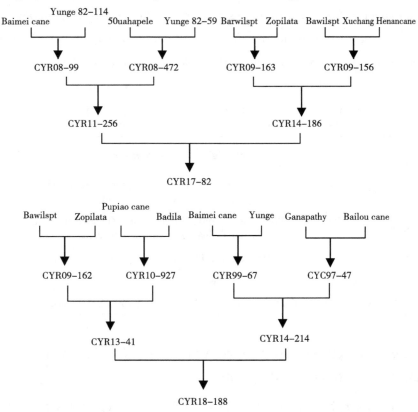

Figure 5-31 The pedigree of CYR17-82 and CYR18-188

Compared with POJ2878 and Co290 parent systems, the advantage of the new parent system was that all the hybrid parent systems were new original species, and all-new parent systems used were 21, which was more than the sum of the two-parent systems of "POJ" and "Co". At the same time, the problem of POJ2878 parent crossover was overcome, and the poor symmetry of Co290 hybridization was avoided.

The results of field experiment observation and hybrid progeny showed that these parents had high yield, high sugar content, strong Ratoon ability, excellent disease resistance, and wide adaptability, and the main character was good heritability. It can be used as parents to select hybrid combinations with existing POJ2878 and Co290 parent systems and offspring-derived varieties. Due to no consanguinity crossover, strong heterogeneity, and obvious heterosis, the parents of the new independent parent system are expected to become the main source for breeding breakthrough sugarcane varieties in China and worldwide (Wu, 2021). The first batch of new parents with F_2 generation of independent kindred has been provided to the Hainan Sugarcane Breeding Station, China, in 2019. The catalog is shown in Table 5-8. After the study of flowering characteristics is completed, the next step is expected to gradually provide the fuzz of the new hybrid progeny to sugarcane breeding organizations nationwide.

Table 5-8 The list of parents to the Hainan Sugarcane Breeding Station in 2019

NO.	Parents and sources	type	quantity
1	CYR15-55(CYR09-170×CYZ03-232), CYR14-168(CYC58-47×CYR09-173), CYR14-178(CYR 09-160×CYR09-16), CYR14-211(CYR09-67×CYC97-47), CYR13-47(CYR 09-169×CYR10-915)	F_2 generation by peer-to-peer hybrid	5
2	CYR15-90(CP94-1100×CYR11-103), CYZ16-1001 and CYZ16-1002(CYT93-159×CYR10-688), CYZ16-1005, CYZ16-1006, and CYZ16-1007(CYR13-46×SP80-0185), CYZ16-1008(CYR14-192×CYZ05-49)	Containing the F_2 generation consanguine by peer-to-peer hybrid	7

5.5.4 Problems in the 'CYZ' and 'CYR' Independent Parents System of Peer-to-peer Hybridization

The process of sugarcane crossbreeding is the utilization of heterosis, which refers to the phenomenon that F_1 generation hybrids produced by the cross of two parents with different traits are significantly superior to their parents in yield, quality, disease resistance, adaptability, and other aspects (Wu, 2005, 2018, 2021; Wu et al., 2013). Parents with more excellent traits, less cross kinship, strong heterogeneity, and good combining ability will have more excellent offspring, and it is easier to breed breakthrough varieties (Wu, 2005, 2018, and 2021; Wu et al., 2013).

The practice has proved that parents with independent consanguinity have good genetic characteristics and outstanding super-parent advantages and can breed breakthrough sugarcane varieties (Wu, 2005, 2018, and 2021; Wu et al., 2013).

The main problems existing in the parent system of breeding new sugarcane independent consanguinity through Peer-to-peer hybridization in Yunnan are as follows:

Firstly, the most important sugarcane traits with large diameter, juicy, low fiber, sugar content, and resistance to smut are *S. officinarum* L., whose origin centers are located in eastern Indonesia and New Guinea of the Pacific Ocean (Luo, 1984; Li et al., 2010; Chen et al., 2011; Wu, 2005; Wu et al., 2013; Wu, 2018; Zhuang et al., 2013; Wu et al., 2014). Few *S. officinarum* L. are preserved in the National Germplasm Repository of Sugarcane, China, and *S. officinarum* L. resources are scarce.

Secondly, the number of other cultivated original species (including *S. sinense* Roxb., *S. barberi*, and endemic species) collected by the National Germplasm Repository of Sugarcane is less too, In particular, the number of original species with good traits performance.

Thirdly, it is difficult for the original species to blossom, flowering synchronization, and the success rate of hybridization is low. Creating new independent parents makes it easy to make sure that all the original species can be effectively used as we designed. The original species of the same parent do not repeat. The traits are completely complementary in selecting and matching. For example, Bawilspt, one of *S. officinarum* L. in CYR 17-82 parents, was used twice. At the same time, it also contains the consanguinity of Yunnan 82-114 and Yunnan 82-59 (*S. spontaneum* L.), resulting in the consanguinity of wild species as high as 25%.

Fourly, some parents contain the relatives of *E. arundinaceum* (Tao et al., 2014; Tao et al., 1997). E. arundinaceum, as a relative plant of sugarcane, has not been bred into good varieties after decades of cross-utilization at home and abroad, and the effect of improving sugarcane varieties needs to be further verified.

Finally, in theory, 8 original varieties are needed to breed a good new independent consanguineous F_3 parent, among which at least 7 original cultivated species (including *S. officinarum* L., *S. sinense*, *S. barberi*, and Landrace) can be used as the female parent. There are only 16 original cultivated species in use at 'CYR' and 'CYZ' present, and at most, two parental systems with completely different blood ties can be created. Therefore, blood crossing among different parents is inevitable. For example, CYR17-82 and CYR18-188 have no duplication with the two independent parent systems of "POJ" and "Co", but they have the same four parent systems of Baimei cane, Yunnan82-114, Barwilspt, and Zopilata (Figure 5-31).

5.5.5 Suggestions on the Cultivation and Utilization of Newly Independent Parents

5.5.5.1 The introduction and utilization of new fine *S. officinarum* L. should be strengthened

Sugarcane (*Saccharum* L.) is native to the tropics and subtropics, and its cultivation and distribution are mainly in those regions. The distribution area is mainly in the 10–30° north-south latitude. The area with the most concentrated production and high yield is about 23.5° north and south latitude, two sides of the Tropic of

Capricorn and the Tropic of Cancer.

However, with the continuous successful hybridization and backcross utilization of wild resources, the sugarcane planting area has been extended to 38° N (Spain) and 33° S (Australia). China has also reached 33° N (Hanzhong, Shanxi), close to the edge of the northern boundary (Wu et al., 2014).

In Yuanjiang, Shuangjiang, and Shiping counties of Yunnan Province, many sugarcane areas are distributed on the plateau with an altitude of 1,400–1,600 m, and some sugarcane areas even reach 2,000 m. In these high-altitude sugarcane areas, sugarcane still shows good agronomic traits (Wu et al., 2014). However, due to the insufficient quantity and low utilization rate of *S. officinarum*, the speed of improvement of important traits such as yield and sugar content in sugarcane varieties was much lower than that of other crops. The quantity, quality, and utilization of *S. officinarum* with excellent traits are the key to creating a new independent parent system with better quality and the cultivation of breakthrough varieties with an independent parent system.

5.5.5.2 Attach great importance to the use of Landrace

In the process of sugarcane hybrid utilization, we need to analyze the relationship between the Landrace and the original cultivated species with the help of modern molecular biological technology (Wu, 2005, 2018, and 2021; Wu et al., 2013). If the landrace is tropical species, it should be used according to the independent parent system of peer-to-peer hybridization. Suppose the local race is the hybrid offspring of the original cultivated species and does not have a blood relationship with "POJ" and "Co". In that case, it can cross with the F_1 generation of the original species, and the original parents can be added according to the idea of reciprocal hybridization to develop a new independent parent system (Wu, 2005, 2018, and 2021; Wu et al., 2013).

Although there is a blood cross with "POJ" and "Co", it has been accepted in production regarding adaptability, disease resistance, and main traits due to a long period of unhybridization and long-term terrogenetic domestication. It can be used as an improved parent, and "sub-breakthrough" varieties may also be bred (Wu, 2005, 2018, and 2021; Wu et al., 2013).

5.5.5.3 Proper use of good parents with new independent blood systems

The existing parents of "POJ" and "Co" systems have been used in high frequency for more than 100 years, and their excellent characteristics have been exhausted, making it more difficult to breed better varieties and lengthen the breeding cycle (Wu, 2005,2018, and 2021; Wu et al., 2013). There is no blood cross between the excellent parents with new independent blood systems and the parents with "POJ" and "Co" systems, with strong heterogeneity and obvious heterosis, which is easy to breed breakthrough varieties. Therefore, it is suggested to increase the intensity of selection between the excellent parents with new independent blood systems and the existing parents (Wu, 2005, 2018, and 2021; Wu et al., 2013). Excellent parents with a new independent consanguinity system have no consanguinity crossing with "POJ" and "Co" parents, strong heterogeneity, and obvious heterosis, which are easy to breed breakthrough varieties. Therefore, increasing the mating intensity between new excellent parents with independent consanguinity system and existing sugarcane parents (Wu, 2005, 2018, and 2021; Wu et al., 2013). However, in selecting specific combinations, we need to analyze the consanguinity composition carefully and avoid using the parents containing the consanguinity of *E. arundinaceum* to cultivate new varieties. After all, no excellent varieties are bred containing the consanguinity of *E. arundinaceum* at home and abroad; Do not select the cross combinations between new independent blood parents because of the limitation of the number of original species easy to cause new blood cross.

References

AN R D, TAO L A, YANG L H, et al., 2008. The fuzzy comprehensive evaluation for new sugarcane clones of Yunrui 04 series[J]. Sugarcane and Canesugar (1): 1-5.

BIAN X, DONG L H, SUN Y F, et al., 2014. Principal component analysis of drought resistance related traits of *Saccharum spontaneum* L. and its F_1 hybrids[J]. Agricultural Research in the Arid Areas, 32 (3): 56-61.

BIAN X, JING Y F, TAO L A, et al., 2015. Difference-similarity analysis of hybrid combinations with wild sugarcane blood in Yunnan[J]. Sugar Crop of China, 37 (2): 29-31.

CHEN R K, XU L P, LIN Y Q, et al., 2011. Modern sugarcane genetic breeding[M]. Beijing: China Agriculture Press (in Chinese).

CHU L B, 2000. "YN" Study on sugarcane breeding system-application "Heterogeneous composite separation theory" A super - superior germplasm of Yunnan *Saccharum Spontaneum* L. F_1 with high sugar content was obtained[J]. Sugarcane (4): 22-33.

DENG H H, LI Q W, CHEN Z Y, 2004. Breeding and utilization of new sugarcane parents[J]. Sugarcane, 11(3): 7-12.

FAN Y H, CHEN H, SHI X W, et al. 2001. RAPD Analysis of *Saccharum spontaneum* from different ecospecific colonies in Yunnan[J]. Acta Botanica Yunnanica, 23 (3): 298-308.

HU X, XIA H M, YANG L H, et al., 2021. Identification and evaluation the resistance to smut in sugarcane parental clones[J]. Sugar Crops of China, 43(1): 45-50.

IRVINE J E, 1999. *Saccharum* species as horticultural classes[J]. Theoretical and Applied Genetics (98): 186-194.

JING Y F, BIAN X, TAO L A, et al., 2014. Factor and cluster analysis of Yunnan innovated germplasm materials F_1 of *Spontaneum*[J]. Journal of Plant Genetic Resources, 15 (1): 177-181.

JING Y F, CHU L B, TAO L A, et al., 2008. The study on ratooning ability of the new sugarcane varieties/clones breeding with consanguinity *S. spontaneum*[J]. Sugar Crop of China (2): 4-7.

JING Y F, DONG L H, SHUN Y F, et al., 2013. Genetic analysis of drought resistance of different ecotypes in Yunnan[J]. Journal of Hunan Agricultural University, 39 (1): 1-6.

LANG R B, JING Y F, AN R D, et al., 2015. Evaluation by DTOPSIS on cross combinations from Yunrui innovative parent lines and introduced sugarcane lines[J]. Sugar Crops of China, 37 (5): 43-45.

LI Q W, CHEN Z Y, LIANG H, 2000. Modern technology for sugarcane improvement[M]. Guangzhou: South China University of Technology Press.

LI Q W, DENG H H, ZHOU Y H, et al., 2000. Recent Studies on Flowering Induction and Utilization of New Genetic Resources in Sugarcane at Hainan Sugarcane Breeding Station[J]. Sugarcane and Canesugar, 1:1-7.

LI Y R, 2010. Modern sugarcane science[M]. Beijing: China Agriculture Press (in Chinese).

LIU J Y, CHEN X K, FU J F, et al., 2003. Analysis on the genetic relationship of several fine sugarcane lines[J]. Sugar Crops of China (2):1-5.

LIU X L, SU H S, LIU H B, et al., 2014. Correlation and clustering relationship analysis of Yunnan octoploid clones of *Saccharum spontaneum* in China on the basis of yield and quality related traits[J]. Southwest China Journal of Agricultural Sciences, 27 (4): 1382-1386.

LUO J S, 1984. sugarcane science[C]. Guangzhou: Guangdong Sugarcane Society (in Chinese).

MIRJKAR S J, DEVARUMATH R M, NIKAM A A, et al., 2019. Sugarcane (*Saccharum* spp.): breeding and genomics[M]// AL-KHAYRI J M. Advances in plant breeding strategies: industrial and food crops. 6. ed. Switzerland: Springer, 11: 363-406.

Peng S G. 1990. Sugarcane Beeding[M]. Beijing: China Agriculture Press (in Chinese).

TAO L A, DONG L H, JING Y F, et al., 2014. Compositive evaluating of the grey closeness degree for the hybrid F_1 of the *Saccharum* species[J]. Journal of Plant

Genetic Resources, 15 (6): 1248-1254.

TAO L A, JING Y F, DONG L H, et al., 2020. Integrated evaluation of the YR15 sugarcane innovative germplasm with different ways of crossing by DTOPSIS[J]. Sugar Crops of China, 42 (1): 13-21.

TAO L A, JING Y F, DONG L H, et al., 2011. Genetic analysis of main traits in descendants of crossing with *Saccharum spontaneum* 82-114 in Yunnan[J]. Journal of Plant Genetic Resources, 12 (3): 419-424.

TAO L A, YANG L H, AN R D, et al., 2015. Comprehensive analysis and comparison of five methods for 08 series *Saccharum spontaneum* F_2 in Yunnan[J]. Southwest China Journal of Agricultural Sciences, 28 (5): 1907-1915.

TAO L A, YANG L H, JING Y F, et al., 2011. Comprehensive evaluation of drought resistance for descendant F_2 of *Saccharum spotaneum* in Yunnan by subjection function[J]. Southwest China Journal of Agricultural Sciences, 24 (5): 1676-1680.

TAO L A, ZHANG J R, 1997. A preliminary study on the characteristics of S. *spontaneum* L. and E. *arundinaceum* F_1-generation superior materials with resistance to chopping and sugar transformation[J]. Sugarcane (2): 9-11.

TIAN C Y, TAO L A, YU H X, et al., 2017. Drought resistance evolution of F_1 and F_2 hybrids from five climatic ecotypes *Saccharum spontaneum* L. [J]. Agricultural Sciences in China, 50 (22): 4408-4421.

WU C W, 2002. Analysis of the utilization and efficiency of the parents for sugarcane sexual hybridization in Yunnan[J]. Sugarcane and Canesugar (4):1-5.

WU C W, 2005. Discussion on Germplasm Innovation and Breeding Breakthrough Varieties in Sugarcane[J]. Southwest China Journal of Agricultural Sciences, 17 (6): 858-861.

WU C W, 2018. Discussion on breakthrough of inbreeding and interleaved use of sugarcane parents in creating new sugarcane parents. Sugar Crops Improvement, Biotechnology, Bio-Refinery, and Diversification: Impacts on Bio-based Economy[C]. Udon Thani, Thailand, 407-410.

WU C W, 2021. Advances in breeding new independent parents of sugarcane by the peer to peer hybridization in Yunnan[J]. Sugar Crops of China, 43(3): 37-41.

WU C W, ZHAO J, LIU J Y, et al., 2014. Modern sugarcane seed industry[M]. Beijing: China Agriculture Press (in Chinese).

WU C W, ZHAO P F, XIA H M, et al., 2014. Modern cross breeding and selection techniques in sugarcane[M]. Beijing: Science Press (in Chinese).

YU H X, JING Y F, LANG Y B, et al., 2015. Evaluation on new sugarcane strains Yunrui07 series by DTOPSIS[J]. Sugar Crops of China, 37(4): 33-35.

YU H X, TAO L A, TIAN C Y, et al., 2019a. Utilization of *S. robustum* 57NG208 in breeding of YR-series parental clones[J]. Sugar Crops of China, 41(2): 1-7.

YU H X, TIAN C Y, JING Y F, 2019b. Principal component clustering analysis and evaluation of F_2 generation from Yunnan *Saccharum sponta*neum L. innovation germplasm[J]. Journal of Plant Genetic Resources, 20 (3): 624-633.

ZHANG M Q, WANG H Z, BAI C, et al., 2006. Genetic improvement and efficient breeding of sugar crops[J]. Beijing: China Agriculture Press (in Chinese).